Genetics,

Diversity, and the Biosphere

Denson K. McLain, Ph.D.
Georgia Southern University

Other AudioText publications:

Biology-On-Tape: *Understanding the Cell*
Biology-On-Tape: *Genetics, Diversity, and the Biosphere*
Biology-On-Tape: *Human Anatomy & Physiology - Part I*
Biology-On-Tape: *Human Anatomy & Physiology - Part II*

Understanding the Cell
Human Anatomy & Physiology - Part I
Human Anatomy & Physiology - Part II

Printed in the United States of America
First Edition

AudioText, Inc.
P.O. Box 690791
Houston, TX 77269-0791

International Standard Book Number:
1-884612-03-2

Library of Congress Catalog Card Number:
94-72454

1 2 3 4 5 6 7 8 9 - 99 98 97 96 95

TABLE OF CONTENTS

CHAPTER 1

NUCLEIC ACIDS & CHROMOSOMES

Genetics is the study of the patterns and mechanisms of inheritance. Patterns of inheritance were first elucidated in the 19th century, by the Austrian monk, **Gregor Mendel**, who's considered to be the father of genetics. Mendel studied the inheritance of traits such as flower color, seed shape, height, and pod color in the common pea. Remarkably, Mendel discerned principles regarding the patterns of inheritance by following the transmission of traits between generations, without any knowledge of the molecular mechanisms of inheritance nor, indeed, without any knowledge that DNA is the hereditary material.

In the 20th century double-stranded DNA was established as the hereditary material of all organisms, whether bacterium, protozoan, fungus, plant or animal. The only exceptions are some viruses that employ either RNA or single-stranded DNA as their genetic material. Still, most viruses use DNA. In the mid-20th century **James Watson** and **Francis Crick** deduced the structure of DNA. Since then, our understanding of the molecular basis of inheritance and our capacity to precisely manipulate the genetic constitution of organisms has grown at an accelerating pace.

We'll discuss aspects of modern biotechnology soon. But, first, it's necessary to develop the proper background. We'll describe the structure of DNA and chromosomes first. Next, we'll discuss the cell cycle, including mitosis and meiosis. Then we'll explore Mendelian inheritance before returning to molecular mechanisms and genetic engineering. **Key words** for our discussion of DNA structure are: nucleotide, deoxyribose, base, and antiparallel.

DNA consists of two molecules, or strands, wound around each other. Each strand contains thousands to tens of millions of building blocks called **nucleotides**, all arranged one-behind-the-other in a long line. Each nucleotide has three parts: a sugar, a phosphate group, and a base. The sugar is called **deoxyribose**. Its skeleton consists of four carbons and an oxygen that form a 5-membered ring. These carbons are numbered 1' to 4'. The oxygen atom lies between the 1' and 4' carbons. A fifth carbon, the 5' carbon, is also attached to the 4' carbon. Although the 5' carbon is attached to the ring, it's not a member of the ring.

The **phosphate group**, PO_4, is attached to the 5' carbon. Another molecule, the base, is attached to the 1' carbon. Picture the nucleotide building block as L-shaped. The deoxyribose sugar is the corner of the L, the phosphate is the vertical part of the L, and the base is the horizontal part. To make a strand of DNA stack the Ls, one on top of the other. The resultant structure is comb-shaped; the bases stick out, all on the same side, like the teeth of a comb. The backbone of the comb is made of the vertical parts of the stacked Ls which correspond to the deoxyribose sugar and phosphate. So, the DNA strand is a backbone that starts out with a sugar, then reads phosphate, sugar, phosphate, sugar, phosphate, sugar, *etc.*, until it terminates with a phosphate.

The end of the strand that terminates with a phosphate group is called the 5' end while the end that terminates with a sugar is called the 3' end. This is because the phosphate is attached to the 5' carbon of its nucleotide and attaches or binds to the 3' carbon of the deoxyribose of the nucleotide above it.

Keeping our analogy of stacked Ls, now imagine two strands of DNA, one beside and to the right of the other. The bases face the same direction, to the right. If we turn the right strand upside down, its bases now face the bases of the strand to its left. Note, however, that the 5' end of one strand is opposite the 3' end of the other strand. That is, the strands are **antipar-**

allel. Antiparallel means the strands lie beside each other but are oriented in opposite directions.

STRUCTURES OF RNA & DNA

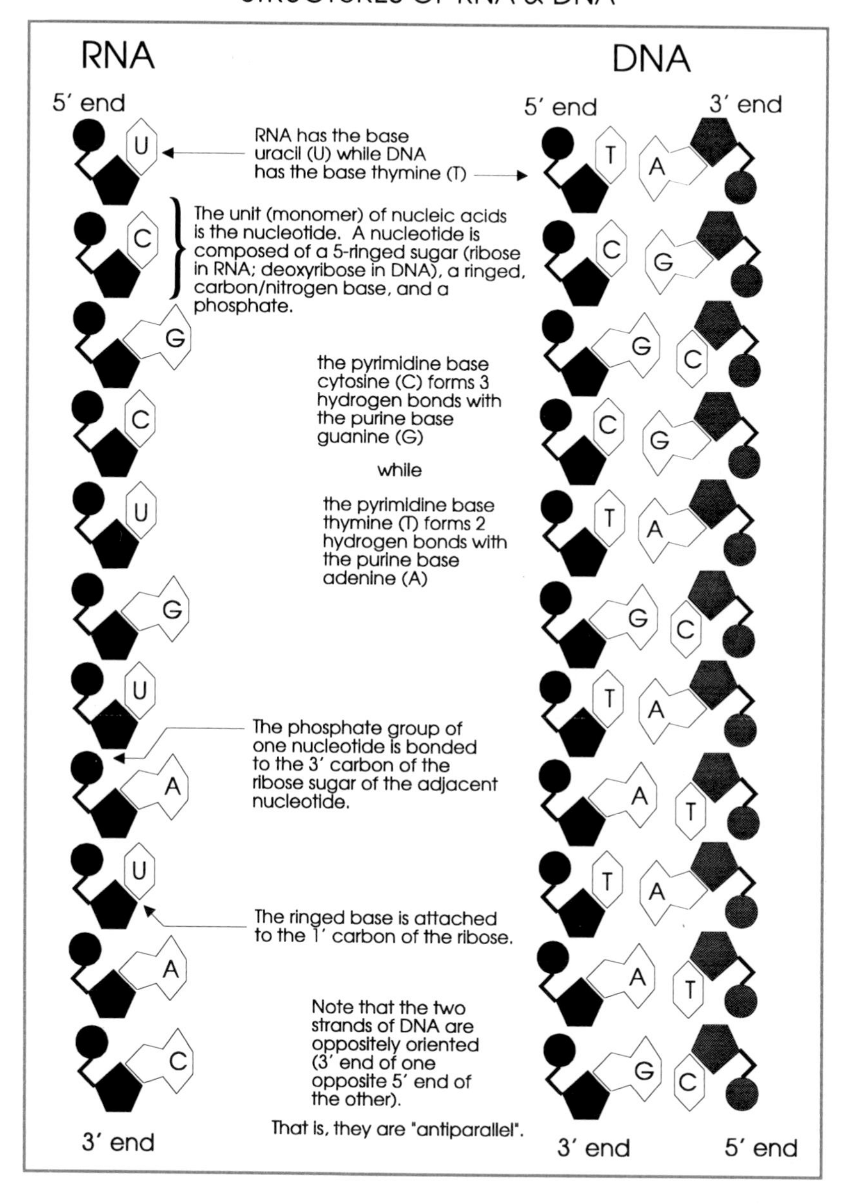

If we push together these two antiparallel strands of stacked Ls, the bases make contact and the structure resembles a ladder, just as holding two combs together would do. It's the interaction between bases in the form of **hydrogen bonds** that holds the two strands of DNA together. Here, a hydrogen bond occurs when a nitrogen atom in one base and an oxygen atom in another base are attracted to, and share, the same hydrogen atom.

Bases are small molecules with a skeleton composed of either a single 6-membered nitrogen-carbon ring or two attached nitrogen- carbon rings, one ring with 6 atoms and one with 5. Bases with a single ring are called **pyrimidines**. There are two pyrimidines in DNA: **cytosine**, symbolized by C, and **thymine**, symbolized by T. Bases with two rings are called **purines**. There are two purines in DNA: **adenine**, symbolized by A, and **guanine**, symbolized by G.

Hydrogen bonding between bases always involves one purine and one pyrimidine. In this way the distance between the two DNA strands stays almost constant because purines are twice as large as pyrimidines. The purine A always bonds with the pyrimidine T while the purine G always bonds with the pyrimidine C. A and T form two hydrogen bonds between themselves while G and C form three.

COMPLEMENTARY DNA BASES

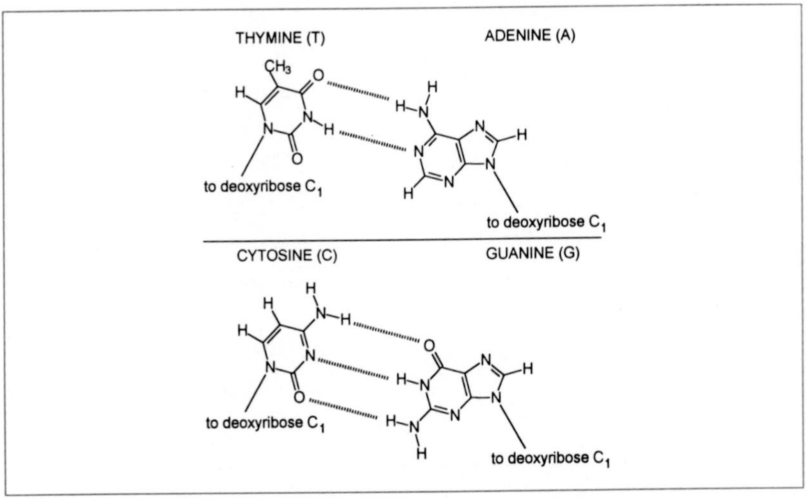

THYMINE (T) ADENINE (A)

to deoxyribose C_1 to deoxyribose C_1

CYTOSINE (C) GUANINE (G)

to deoxyribose C_1 to deoxyribose C_1

Now that we know how bases hold the two antiparallel strands together, let's complete our picture of DNA. Recall that stacked Ls give rise to a comb; two combs give rise to a ladder. In the ladder analogy, the rungs of the ladder are the pairs of bases on alternate strands held together by hydrogen bonds. Now imagine a ladder held in place at its bottom and twisted counter-clockwise at its top. In this case, the rails of the ladder coil about each other somewhat like the railings of a spiral staircase. The coiling is referred to as a **helix**; hence the synonym for DNA, the double helix. There are about 10 pairs of bases per complete turn of the double helix.

Some viruses use **RNA** as their hereditary material. RNA is a single-stranded molecule. It's composed of nucleotides, and is similar to a single strand of DNA. But, two features distinguish an RNA strand from a single strand of DNA. First, RNA has **ribose** instead of deoxyribose in its backbone. Ribose is just like deoxyribose except that at the 2' carbon a hydoxyl, or OH, group replaces a hydrogen atom. Second, RNA substitutes the base **uracil**, symbolized by U, for thymine.

Now let's **review the key words**: nucleotide and deoxyribose. DNA is composed of two strands wound around each other. Each strand is a string of thousands to millions of units called nucleotides. Each nucleotide is composed of a central sugar molecule, deoxyribose, to which is attached a phosphate group and a ringed molecule called a base. Deoxyribose has five carbon atoms, four of which, along with an oxygen atom, occur in a five-sided ring. Phosphate is attached to the 5' carbon which is not a part of the ring while the base is attached to the 1' carbon. The "deoxy-" part of deoxyribose refers to the replacement of a hydroxyl or OH group at the 2' carbon of ribose with a hydrogen atom. RNAs have ribose, not deoxyribose, in their backbones.

Now let's **review the key words**: base and antiparallel. There are four different bases in DNA, two purines, guanine, G, and adenine, A, and two pyrimidines, cytosine, C, and thymine, T. The two strands of DNA are held together by hydrogen bonds between guanine on one strand and cytosine on the other and between adenine on one strand and thymine on the other. The two hydrogen bonded strands of DNA are oriented in opposite directions. That is, they are antiparallel. The 5' end of one strand is opposite the 3' end of the other strand.

Now let's discuss the distribution of genes in DNA and describe the structure of chromosomes. **Key words** for this section are: gene, allele, locus, chromosome, nucleosome and histone.

DNA is called the blueprint for life because it holds the information used to build cells and bodies. Sequences of the four bases, A, G, C, and T, hold this information. Just as letters hold information when they are organized into words and sentences, so too can base sequences hold information. And like words, base sequences are said to be read. Reading the information content of DNA really means unzipping the hydrogen bonded strands of the double helix and using one strand as a guide or **template** to make an RNA copy of the other strand. The RNA duplicate or **transcript** is made by linking together nucleotides as their bases hydrogen bond with the template according to the rule U pairs with A and G pairs with C.

Gene expression entails both transcription and translation. **Transcription** is the process by which the information of DNA genes is copied into RNAs. DNA-associated enzymes, called **RNA polymerases**, catalyze and direct the production of transcripts. The three most important classes of transcripts are **ribosomal RNAs**, **transfer RNAs**, and **messenger RNAs**. These interact at a **ribosome** to produce the particular protein product specified in the messenger RNA. **Translation** is the process of reading messenger RNA and using that information to produce proteins.

A **gene** is a relatively short, specific sequence of DNA read in the process of transcription. All gene transcripts function in some aspect of protein synthesis. Therefore, the inheritance of traits that results when parents pass on copies of their genes to their offspring is ultimately a reflection of protein synthesis.

Let's consider an example. Some strains of white clover synthesize the poison hydrogen cyanide, or HCN, as a defense against herbivores. This small chemical is not a protein but its production depends on protein synthesis. The final step of the metabolic pathway that leads to HCN production requires an enzyme, or catalytic protein, to liberate hydrogen cyanide from another chemical called cyanogenic glucoside. The enzyme required for this last step is made by the translation of messenger RNA.

RNA SYNTHESIS: TRANSCRIPTION

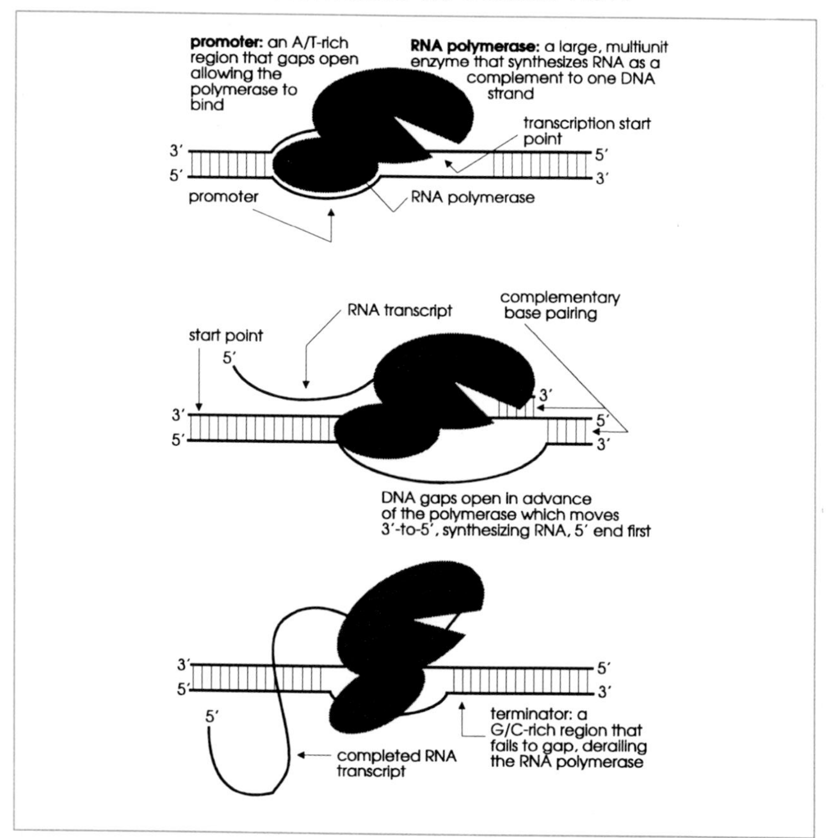

Different strains of clover have different sequences in their gene for this enzyme. Consequently, they give rise to different messenger RNAs which are then translated into slightly different forms of the enzyme. Some forms of the enzyme are inactive and, therefore, cannot convert glucoside into hydrogen cyanide.

Inheritance of HCN production, then, depends on which form of the gene parents pass to their offspring. Which gene is received from the parents, then, determines which enzyme is produced which, in turn, determines if HCN is produced. HCN is not itself a protein but, as a trait, its expression depends on the type of protein, in this case an enzyme, the offspring can produce.

PROTEIN SYNTHESIS

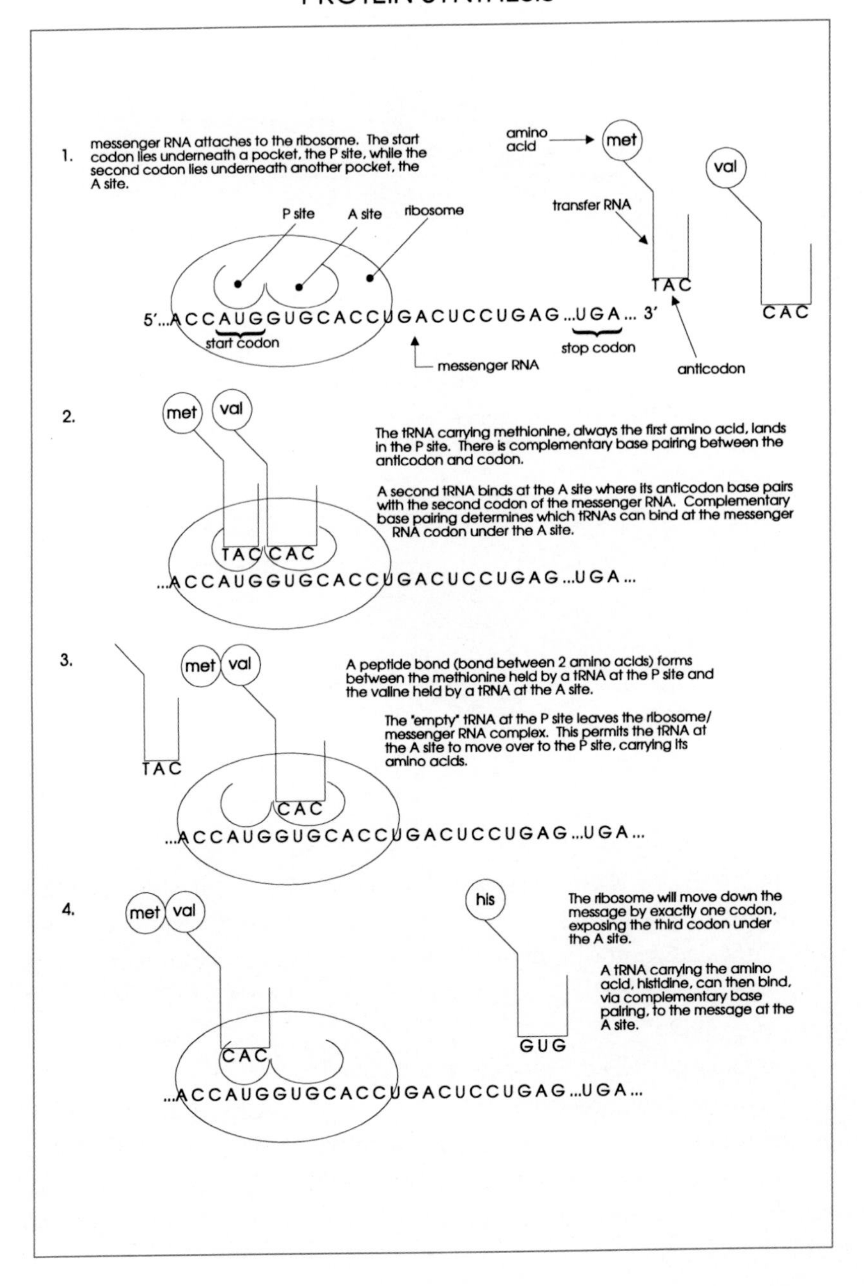

PROTEIN SYNTHESIS (cont.)

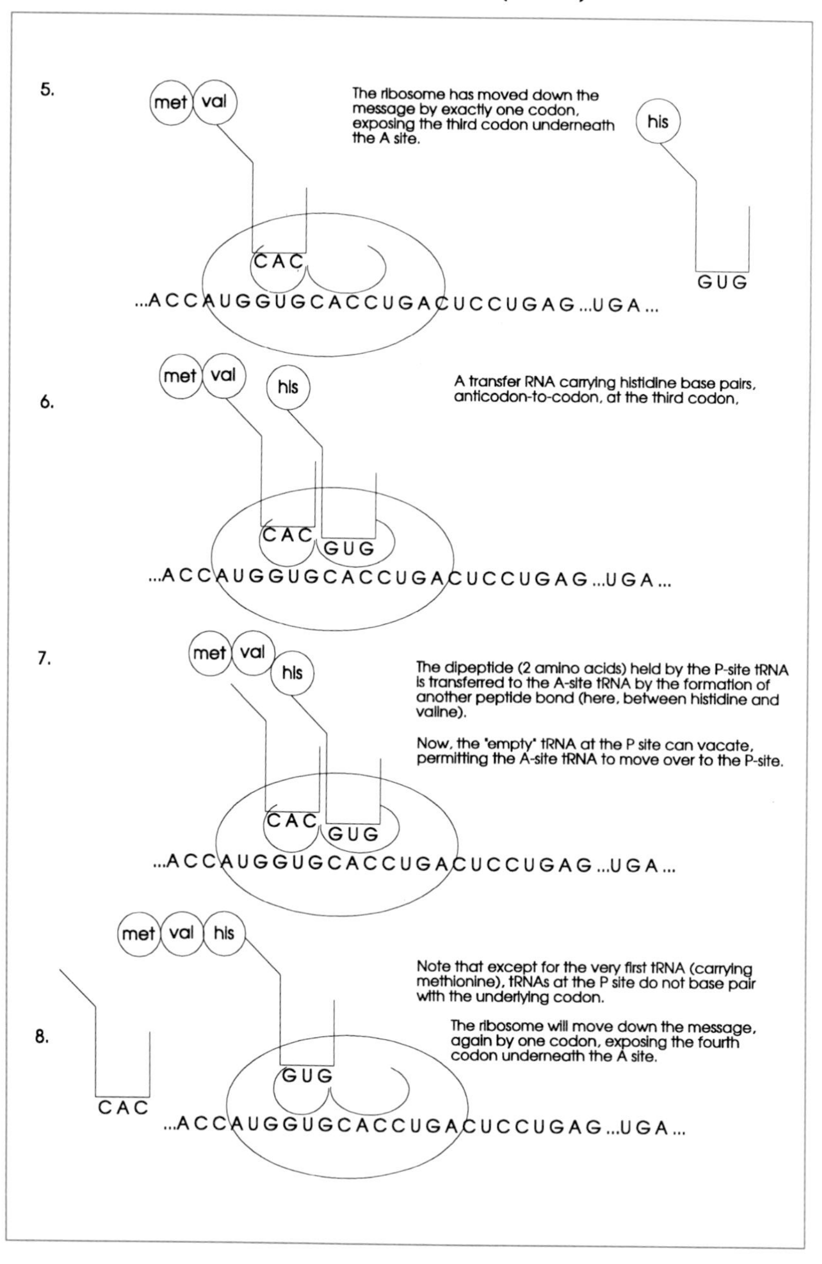

PROTEIN SYNTHESIS (cont.)

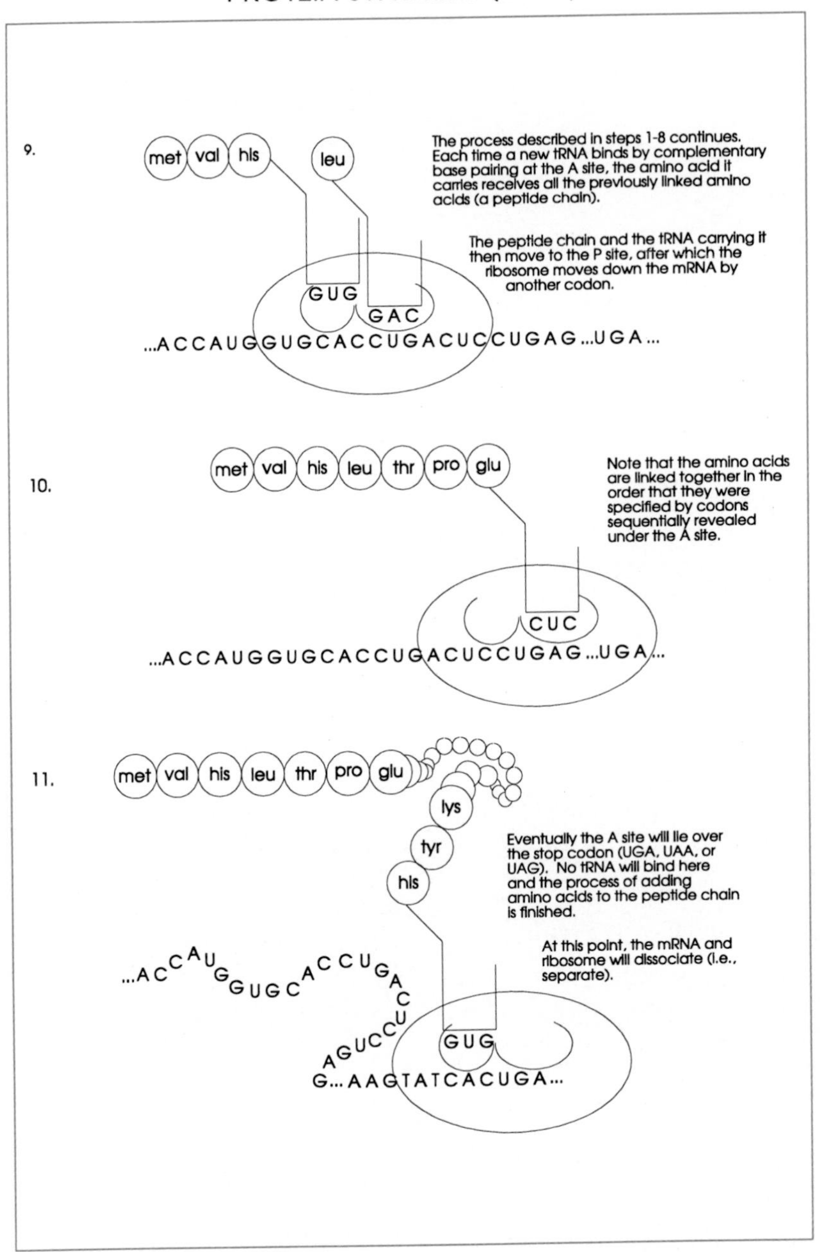

Genes are the hereditary unit passed from generation to generation, from parent to offspring. Different forms of a gene are called **alleles**. For instance, a gene for eye color might have two alleles, one allele causing blue eye color and one allele causing brown eye color. Actual eye color is determined by the combination of alleles an individual possesses. Allelic differences between individuals can cause visible differences between individuals. For instance, we can properly speak of an allele for white flowers and an allele for red flowers in carnations. This just means the differences in flower color are caused by different alleles at the gene for flower color.

QUANTITATIVE OR POLYGENIC TRAITS

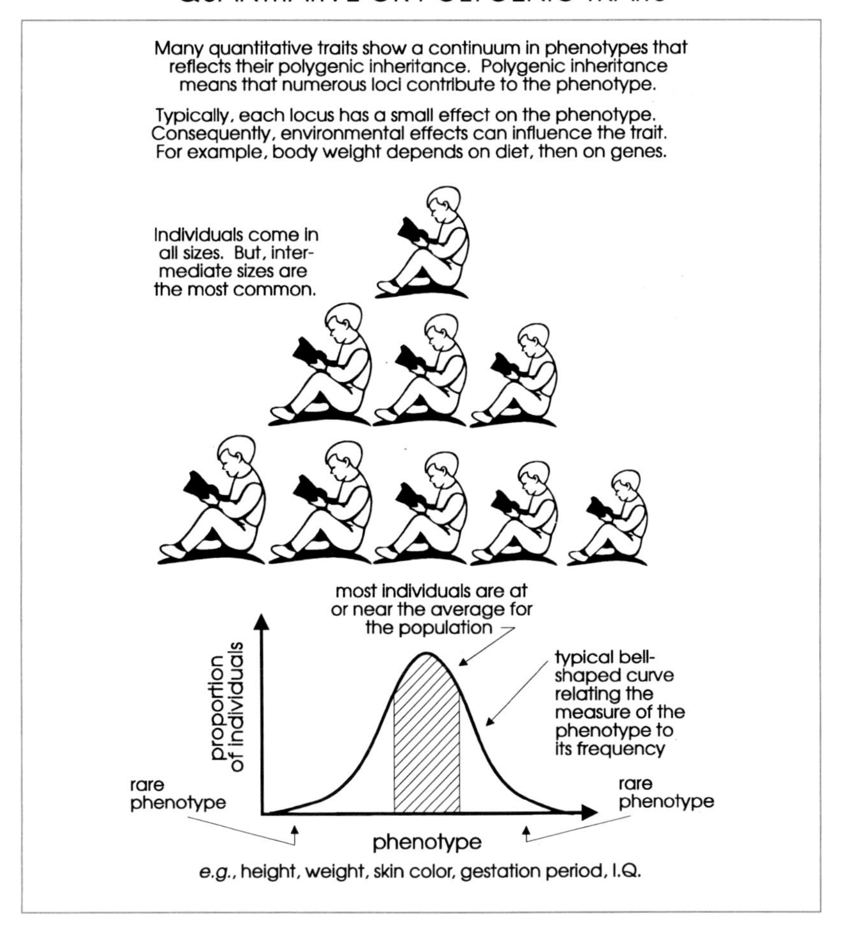

Many traits are determined by the alleles of more than one gene. Each gene that contributes to a trait is a **locus**. The plural of locus is **loci**. Therefore, some traits are determined by multiple loci. Alleles of a single locus determine whether or not ear lobes in humans are attached to the side of the neck or dangle freely. In contrast, several loci determine the exact eye color in humans and many loci determine height in humans. Traits determined by many different genes or loci are called **quantitative traits** and their inheritance is said to be **polygenic**, meaning caused by many genes. The word locus is often used interchangably with the word gene. But, locus has a more precise meaning than gene.

Locus implies a position or place within the DNA of a chromosome. Genes are distributed within the DNA and, therefore, each gene has its own unique location or locus. The DNA of viruses, bacteria, and mitochondria is nearly all genes; where one gene ends another one begins nearby. However, genes are often widely spaced in the DNA of chloroplasts and, especially, in the DNA of the nucleus, where the DNA between genes is **nongenic**.

Structural genes give rise to messenger RNA transcripts that specify the sequence of amino acids in a particular protein. In many animals these genes occur singularly, isolated from each other by stretches of nongenic DNA of a few hundred to a few thousand nucleotides in length. But, in other animals, several structural genes occur close to each other and are also separated from other such assemblages by many thousands of nucleotides of nongenic DNA. The function of nongenic DNA is unknown. It may serve to bind proteins that catalyze and regulate the processes of transcription and DNA replication.

In the cell, DNA is always associated with proteins. Many of the proteins that catalyze transcription and DNA replication are bound to DNA. A **chromosome** is DNA plus its associated proteins. In eukaryotes, the DNA is wound around a complex of proteins called a **nucleosome**. Each nucleosome contains eight **histone proteins**, two copies of each of four different kinds, plus an associated protein, **ubiquitin**. One hundred forty-five base pairs of DNA wind one-and-three-fourths times around each nucleosome. The wound DNA is called a **super helix**. Once the super helical DNA leaves the nucleosome it interacts with yet another histone protein. The nucleosome plus the odd histone organize the chromosomal DNA into segments of 168 base pairs.

CHROMOSOME STRUCTURE

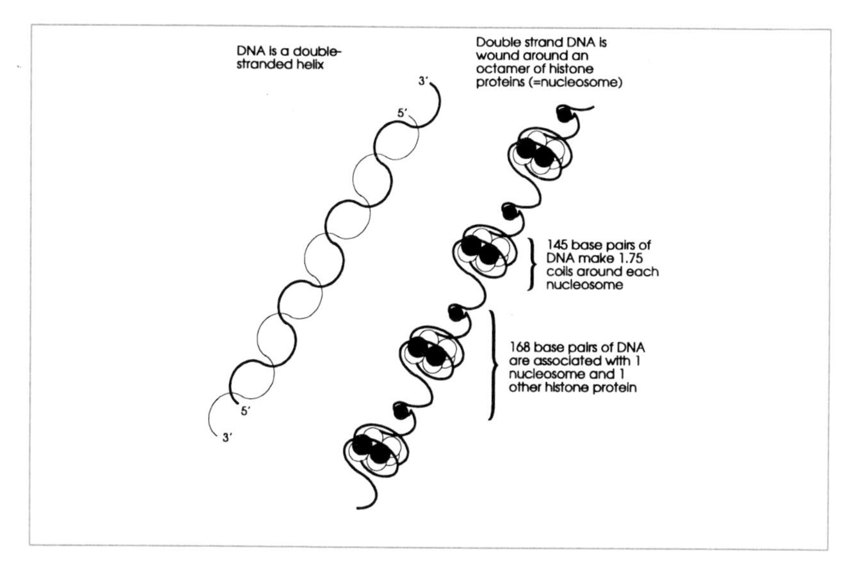

DNA is highly coiled within the nucleus due to its interaction with proteins and cations such as Mg^{2+}. In this way, very long DNA molecules fit within a relatively small nucleus. For instance, a bacterial chromosome composed of four million base pairs of DNA is 1 1/2 millimeters long if not coiled but, when coiled, fits into a bacterial cell almost 1000 times shorter. Likewise, a frog nucleus less than 10 millionths of a meter wide can hold 8 meters of coiled DNA representing about 20 billion base pairs of DNA.

Now let's **review the key words**: gene, allele, and locus. A gene is a stretch of DNA that serves as a guide for the production of a single strand of RNA in a process called transcription. The RNA transcript then serves a role in protein synthesis. Messenger RNAs carry the information that specifies the sequence in which up to 20 different kinds of amino acids are to be linked together to form a protein. Structural genes encode for proteins indirectly by their messenger RNA transcripts. Alleles are different forms of a gene. Alleles of a structural gene might encode or specify slightly different forms of the same protein. For example, an enzyme-encoding gene in white clover has two alleles. One allele gives rise to an active form of the enzyme that synthesizes hydrogen cyanide poison from a chemical precursor while the other allele gives rise to an inactive enzyme that cannot catalyze the synthesis of the poison. A locus is a place on a chromosome where a gene resides. The terms locus and gene are often used interchangeably.

Now let's **review the key words**: chromosome, nucleosome, and histone. A chromosome is DNA plus the proteins associated with it. These proteins include some of the enzymes that catalyze the transcription and replication of DNA. However, the most common proteins associated with DNA are histones. DNA is wound around small complexes of histones called nucleosomes.

CHAPTER 2

MITOSIS & MEIOSIS

A **genome** is all the DNA characteristic of an organism. The genome of humans constitutes about three billion base pairs of DNA, organized into 46 chromosomes within each cell nucleus. Chromosomes can vary in length and in the specific genes distributed along their length. The set of chromosomes characteristic of an organism is called the **karyotype**. How organisms maintain the same genome and karyotype from generation to generation is discussed next. **Key words** for the next section are: haploid and diploid, germ line and soma, and mitosis and meiosis.

Humans, like most plants and animals, possess two copies of each kind of chromosome. Consequently, each genetic locus or gene is represented in the genome by two copies; a condition referred to as **diploidy**. **Diploids** are organisms with two copies of each locus. Diploids receive one copy of each gene from each parent. This is because each parent contributes a complete set of chromosomes to their mutual offspring. A complete set consisting of one of each type of chromosome is called a **chromosome complement**. Cells with just a single chromosome complement are said to be **haploid**.

An integral aspect of sexual reproduction is the formation of haploid **gametes** or sex cells, such as spermatozoa or eggs. In animals, haploid gametes are derived from the diploid **germ-line cells** of a testis, in males, or an ovary, in females. **Meiosis** is the process by which a diploid nucleus gives rise to haploid daughter nuclei. Meiosis shares some similarities with **mitosis**, the process by which a diploid nucleus gives rise to two diploid daughter nuclei.

KARYOTYPE: HUMAN CHROMOSOMES

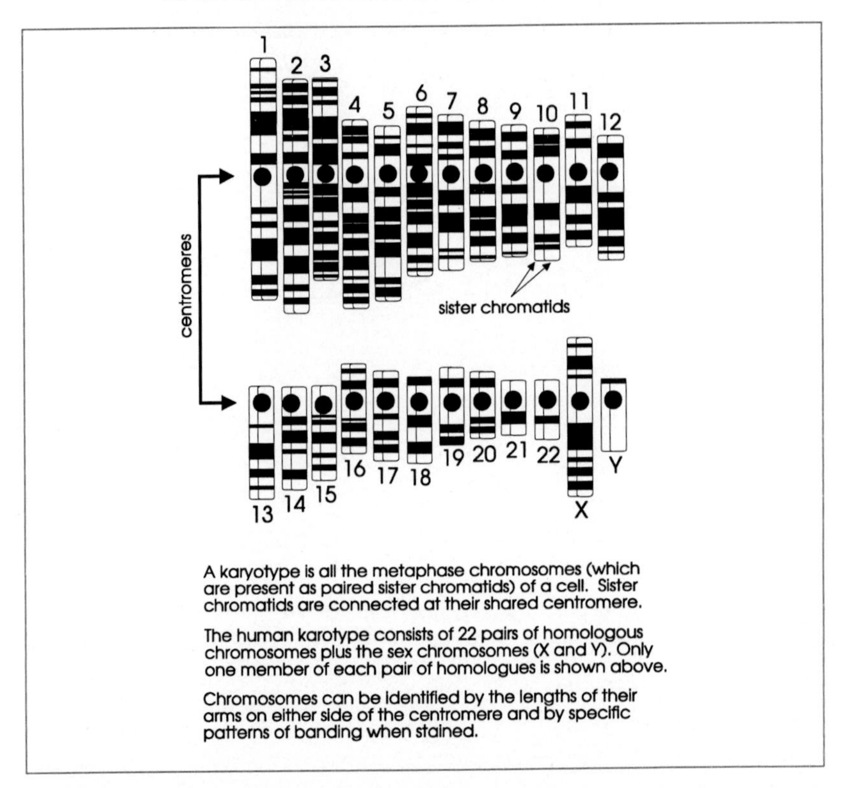

A karyotype is all the metaphase chromosomes (which are present as paired sister chromatids) of a cell. Sister chromatids are connected at their shared centromere.

The human karotype consists of 22 pairs of homologous chromosomes plus the sex chromosomes (X and Y). Only one member of each pair of homologues is shown above.

Chromosomes can be identified by the lengths of their arms on either side of the centromere and by specific patterns of banding when stained.

The cells of an organism can be dichotomized into **somatic** and germ-line cells. Somatic cells are also called body cells; **soma** means body. Somatic cells are diploid. They reproduce during tissue growth and cell replacement. Reproduction of somatic cells entails mitosis. The **germline** is also diploid, and grows by mitosis, but the germline also gives rise to gametes which are haploid. The production of gametes by germ-line cells entails meiosis and is part of the process of sexual reproduction.

Mitosis is a four phase process by which a single diploid nucleus gives rise to two genetically identical diploid nuclei. Prior to the commencement of mitosis, DNA is replicated leading to the duplication of chromosomes. After mitosis is finished cytokinesis may occur. **Cytokinesis** is the division of the cytoplasm. Cytokinesis separates each new nucleus into its own cell.

When chromosomes are duplicated prior to the start of mitosis, the two identical copies, called **sister chromatids**, remain attached to each other at a point called the **centromere**. Attachment at the centromere occurs because DNA replication has not been completed at this point. Consequently, the original complementary strands are still wound around each other at the centromere. Elsewhere, each of the two original strands of DNA is wound around a newly synthesized complement.

The first phase of mitosis is called **prophase**. The DNA becomes tightly coiled in prophase giving rise to shortened and thickened chromosomes. If the centromere is located approximately in the middle of a chromosome, the attached sister chromatids have four arms between them and resemble an X. If the centromere is at one end, the attached sister chromatids have two arms between them and resemble a V. In prophase, the **nuclear membrane** that surrounds the DNA disentegrates. This exposes the DNA to thin protein tubes, called **microtubules**, that bind to each pair of sister chromatids at their centromere. The microtubules radiate out from each pole of the cell and bind to a complex of proteins at the centromere called the **kinetochore**. Each set of radiating microtubules is called a spindle. The **spindle apparatus** is responsible for chromosome movement; and, in animal cells, the spindles arise from barrel-shaped structures called **centrioles**.

The second phase of mitosis is **metaphase**. Here, the pairs of sister chromatids become arrayed across the **spindle equator**. Each spindle apparatus is like the spokes of an umbrella, radiating down in arcs from a central pole. Because the spindles are at opposite ends of the cell, their arcing spokes meet approximately at the middle of the cell in a plane called the spindle equator. The spindle equator is also called the **metaphase plate** to emphasize its planar aspect.

In the next stage, **anaphase**, microtubules that were joined almost end-to-end begin to slide along side each other in a ratchet like process driven by ATP. Consequently, their combined length shortens. As a result, sister chromatids are pulled apart and one of each pair moves toward each **spindle pole**. Since all the chromosomes were duplicated prior to mitosis, one copy of each chromosome originally present moves to each spindle pole.

Telophase is the phase of mitosis when chromosomes actually arrive at their respective spindle poles. Also, nuclear membranes form around each set of chromosomes. Telophase is the last phase of mitosis and is usually followed by cytokinesis, the division of the cytoplasm into two cells.

Mitosis, then, is the somatic cell process by which **daughter nuclei** arise. The daughter nuclei contain identical sets of chromosomes and each is identical to the original nucleus, before its chromosomes were duplicated. The daughter nuclei are now diploid. The number of chromosomes is 2N where N is the number of chromosomes in a complement received from each parent.

Now let's **review the key words**: haploid and diploid, soma and germ line, and mitosis and meiosis. Diploid cells possess two sets, or complements, of chromosomes while haploid cells possess only a single complement. Each complement contains one of each kind of chromosome. Consequently, diploid cells possess two copies of each locus or gene. Body cells, or somatic cells, are diploid whereas gametes, sperm or eggs, are haploid. Gametes are derived from a special tissue called the germ line. The reproduction of diploid somatic cells entails mitosis. Mitosis involves the division of the nucleus following chromosome duplication. Each daughter nucleus receives one copy of each pair of duplicated sister chromatids. The production of gametes from diploid germ line cells entails meiosis. Meiosis consists of two rounds of nuclear division that follow a single episode of chromosome duplication. Consequently, meiosis cuts in half the number of chromosomes present. Each gamete that results from meiosis contains a single copy of each kind of chromosome and is, therefore, haploid.

The number of chromosomes in a diploid cell is said to be 2N while the number in haploid cells is said to be N. N is just the number of different chromosomes.

Meiosis is the process of nuclear division in germ-line cells. **Key words** for our discussion of meiosis are: synapsis, recombination, autosome, homologue, and chiasmata. Meiosis entails a single round of chromosome duplication but two rounds of nuclear division. The first nuclear division restores the diploid number of chromosomes while the second division is reductional. **Reductional division** reduces the number of chromosomes to a single complement. Thus, the resulting cells are haploid.

Meiosis is an integral part of the process of gamete production or **gametogenesis**. In animals, gametes are either egg cells or sperm cells. In sexual reproduction, the penetration of a haploid egg by a haploid sperm leads to **syngamy** or nuclear fusion. Syngamy produces a zygote and restores the diploid or 2N chromosome number.

Meiosis has two functions. First, it's a precise means of reducing ploidy by 1/2 so that each gamete carries a single copy of each type of chromosome rather than two copies of each. Humans have 23 pairs of chromosomes, including the **sex chromosomes**, named **X** and **Y**. Chromosomes other than the sex chromosomes are called **autosomes** or autosomal chromosomes. A human sperm cell contains one **Y** or one **X** chromosome and 22 autosomes while a human egg cell contains one **X** chromosome and 22 autosomes. Each autosome carries a different set of genes.

The second function of meiosis permits the production of genetically diverse progeny by giving rise to genotypically diverse gametes.

The two nuclear divisions of meiosis are called **meiosis I** and **meiosis II**. Each contains the same phases as mitosis; prophase, metaphase, anaphase, and telophase. But, each phase is identified by a roman numeral, I or II, to identify it as part of the first or second meiotic division.

Prophase I is the stage where gene exchange between **homologous chromosomes** occurs. Previously duplicated chromosomes begin to shorten and thicken due to localized coiling at points called **chromomeres**. Due to pairing between the chromomeres of homologous chromosomes, the homologous chromosomes array their arms in parallel. This precise alignment of loci on one chromosome with the same loci on the homologue is called **synapsis**. A frequent consequence of synapsis is **recombination** during which the four parallel chromosome arms become entangled, break, and then reattach in a fashion that results in the mutual exchange of the same genes between homologues. An entanglement of chromosome arms, called **crossing over**, precedes recombination. The X-shaped structure of crossed arms is called a **chiasma** (or chiasmata for plural).

Homologous chromosomes that have mutually exchanged the same loci are called **recombinant chromosomes** or, simply, recombinants. When recombination occurs at a single break point, homologues exchange all of the chromosomal material from the break point to the end of the chromosome arm. When there are two break points, genetic material is exchanged from one break point to the next. In this way, multiple exchanges can occur at various points along a single chromosome arm.

RECOMBINATION

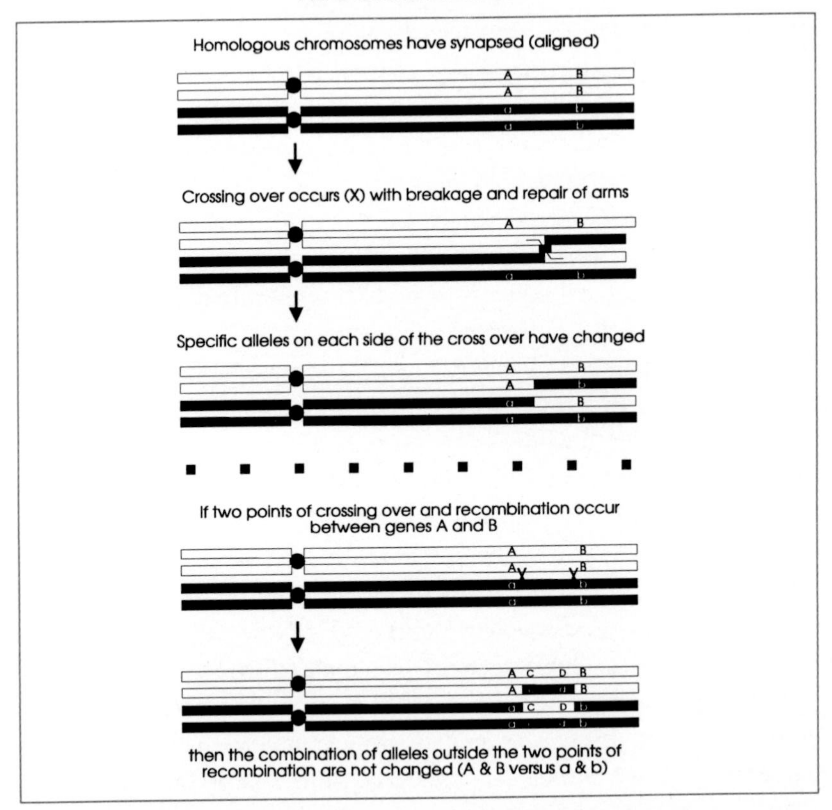

The consequence of recombination is a new collection of nonalleles along a chromosome. Nonalleles are the particular versions of genes at different loci. For instance, suppose the alleles present at two loci on one chromosome are called **A** and **B** while the alleles present on the homologue are **a** and **b**. There are two loci, the **A** locus and the **B** locus. If recombination occurs between the **A** and **B** loci, two new combinations of nonalleles occur. They are **Ab** and **aB** (compare to the original **AB** and **ab**).

The new combination of nonalleles may enhance the survival chances and reproductive success of the zygote relative to zygotes with the original parental combinations of nonalleles. One function of meiosis is to create new combinations of nonalleles. This ability to produce many new combinations of nonalleles each generation is an advantage of sexual reproduction. In a changing environment, this may increase the odds that some offspring will survive and reproduce.

Because synapsis occurs between homologues that have already been duplicated, and because the sister chromatids are still united at their common centromere, synapsis brings together four copies of each chromosome, a pair of sister chromatids for each homologue. These four chromosomes are called a **bivalent**.

Recombination between two loci occurs several ways. When recombination occurs, on average, half of the gametes produced will carry a new recombinant chromosome and half will carry a parental type chromosome because crossing over between any two loci is likely to involve only one member of each pair of sister chromatids. But recombination can also involve all four arms. In addition, double cross-over events can occur. A double cross-over would restore the parental combination of nonalleles on either side of the two points of crossover.

Suppose, for instance, that recombination occurs between the **A** and **B** loci 20% of the time. For those occasions when recombination occurs, 1/2 of the chromosomes are parental types and 1/2 are recombinants. Consequently, 10% of the gametes will carry recombinant chromosomes with regard to the **A** and **B** loci. This calculation is just the rate of recombination, which is 20%, times the proportion of recombinant chromosomes when recombination occurs, which is 50%, or 1/2 the time, in this case.

The rate of recombination between two loci depends upon how far apart they are on a chromosome. The farther apart they are, the higher the rate of recombination due to the greater probability that the synapsed homologues will become entangled at some point between the two loci. When genes are carried on the same chromosome they are said to be **linked**. If the genes are very close together they are tightly linked. Tight linkage between a series of loci reduces the probability that recombination will alter the array of specific nonalleles.

After recombination, the completion of **condensation**, or coiling, marks the end of prophase I. Prophase I is followed by metaphase I. In **metaphase**

I paired homologues align across the metaphase plate, or spindle equator, such that each member of the pair is positioned to move toward opposite spindle poles. Keep in mind that each member of a pair is itself actually a pair of sister chromatids. However, due to recombination the sister chromatids are no longer identical with respect to the nonalleles they carry.

The critical feature of this alignment of homologues is that the set of chromosomes inherited from one parent are not all necessarily positioned to move toward the same spindle pole. Instead, mixtures of maternal and paternal chromosomes move independently toward either spindle pole. Consequently, metaphase I shuffles chromosome complements.

Like recombination, the **shuffling of chromosome complements** is another way meiosis contributes to the genotypic diversity of gametes.

The rest of meiosis I is straightforward. In **anaphase I** the members of each bivalent are pulled to opposite poles. At **telophase I**, the chromosomes have arrived at their respective spindle poles. Cytokinesis, or division of the cytoplasm, then occurs after telophase I. The nuclear membrane may or may not reform prior to cytokinesis. Each cell resulting from meiosis I will most likely have a mixture of maternal and paternal chromosomes but only one member of each original pair of homologues. And each chromosome is present as sister chromatids, still attached at their common centromere.

Meiosis II follows cytokinesis. Meiosis II has all the familiar phases; prophase, metaphase, anaphase, and telophase. The effect of meiosis II is to reduce, by half, the number of chromosomes in the daughter nuclei since there is no DNA replication between meiosis I and II. The reduction is effected by separating the sister chromatids following their alignment at the spindle equator in **metaphase II**. Cytokinesis follows telophase II.

Meiosis results in the production of four haploid cells from one diploid cell. The four cells result from two rounds of cytokinesis. The first follows meiosis I, giving rise to two cells. Each of these two cells then procede with meiosis II and cytokinesis. In **spermatogenesis**, each product of meiosis can become a mature spermatozoan, or sperm cell. In **oogenesis**, only one of the four haploid cells, an especially large cell, becomes the egg or oocyte. The other three cells are **polar bodies** which, in some animals, serve to supply the larger egg with nutrients or even messenger RNA transcripts to facilitate rapid protein synthesis.

Diploid organisms, such as ourselves, receive one complement of chromosomes from the haploid egg, supplied by the mother, and one complement of chromosomes from the haploid sperm, supplied by the father. The union of two haploid cells, sperm and egg, gives rise to a diploid nucleus. Diploidy is then maintained by mitotic divisions as somatic cells proliferate into tissues and organs.

Now let's **review the key words** homologue and autosome. Homologues are a pair of chromosomes that carry the same sets of genes but not, necessarily, the same alleles at each gene. Sexually reproducing diploid organisms receive one member of each pair of homologues from each parent. There are two kinds of chromosomes, autosomes and sex chromosomes. Autosomes exist as pairs of homologous chromosomes. Humans have 23 pairs of chromosomes, 22 pairs of autosomes, each pair being a set of two homologues, and one pair of sex chromosomes. In females, the sex chromosomes are homologous **X** chromosomes whereas, in males, the pair of sex chromosomes, one **X** and one **Y**, are not homologous. The sex with identical sex chromosomes is said to be homogametic while the sex with different sex chromosomes is said to be heterogametic. Females are homogametic in mammals but heterogametic in birds. Some organisms such as many reptiles do not possess sex chromosomes. In crocodiles and aligators sex is determined by temperature.

Now let's **review the key words**: synapsis, recombination, and chiasmata. Synapsis is the precise, locus-by-locus alignment of homologues in prophase I of meiosis. Synapsis gives rise to a bivalent which is a set of four chromosomes representing two homologues, each of which is a pair of attached sister chromatids. Synapsis enables crossing over to occur between arms of homologues. Crossing over is the entangling of arms of homologues, visible as X-shaped structures within the bivalent. An X-shaped point of crossing over is called a chiasma. Crossing over can lead to recombination which is the exchange of genetic material between homologous chromosomes. Recombination can lead to new combinations of nonalleles along a chromosome and is the most important cause of genetic diversity between individuals within a population.

Sometimes errors occur in meiosis that lead to the presence of extra chromosomes in some gametes and missing chromosomes in other gametes.

Nondisjunction refers to the failure of a pair of sister chromatids to separate during meiosis II. For instance, in humans, chromosome 21 sister chromatids may fail to separate. Consequently, one gamete will have two copies of this chromosome instead of the normal single copy while another gamete will have no copies. If a zygote is formed that contains only one chromosome 21, it will spontaneously abort. If, however, a zygote is formed from the fusion of an egg with two copies of chromosome 21 and a sperm with the usual single copy of that chromosome, trisomy 21 occurs. Embryos with **trisomy 21** have three copies of chromosome 21. Children with trisomy 21 suffer from a multitude of afflictions including mental retardation.

Aneuploidy refers to a karyotype in which one to a few chromosomes have extra or missing copies. Aneuploidy, as in the example of trisomy 21, usually results from nondisjunction. Aneuploids resulting from **X** chromosome nondisjunction are **Turner syndrome** females, with the karyotype **X0,** (the zero signifies a missing **X** chromosome) and **Klinefelter syndrome** males, with the karyotype **XXY**. In both instances, the individuals are sterile and lack full development of the appropriate secondary sexual characteristics.

CHAPTER 3

1-LOCUS PATTERNS OF INHERITANCE

We're now ready to understand the basics of transmission genetics which deals with patterns of inheritance. **Key words** for this section are: genotype and phenotype, homozygote and heterozygote, dominance and recessiveness, codominance and incomplete dominance, and sex linkage. Patterns of inheritance are determined by the combination of alleles present at one or more loci within the diploid cell. Diploidy means that each locus is represented by two copies, one received from each parent. Together, the two copies constitute the genotype of that locus. **Genotype** is defined as the combination of alleles present at one or more loci. For instance, consider a locus for which there are two alleles, called **A** and **a**. Three diploid genotypes are possible. Some individuals could be **AA**, some could be **Aa**, and some could be **aa**. A genotype could also be defined with respect to two or more loci. For example, **AaBb** is a genotype at two loci, the **A**-locus and the **B**-locus.

Alleles are different forms of the same gene. A genotype with two identical alleles, such as **aa**, is called **homozygous**. An individual with a homozygous genotype is called a **homozygote**. A genotype of two different alleles, such as **Aa**, is called **heterozygous**. And, correspondingly, an individual with a heterozygous genotype is called a **heterozygote**. Individuals are likely to be heterozygous at some loci and homozygous at others. Therefore, designating an individual as a homo- or heterozygote is always made with reference to a particular gene or locus.

The expression of a gene gives rise to an observable trait, or **phenotype**. Like genotype, phenotype is defined with reference to a particular locus or set of loci. For instance, with respect to genes for flower color and pod color, a pea plant might have this phenotype: purple flowers and green pods. Another pea plant might have the phenotype white flowers and yellow pods.

For diploids, an important consideration arises when the genotype is heterozygous. Is the phenotype a consequence of the expression of both alleles or does one allele take precedence over the other? The simple answer is that it varies from locus to locus and ultimately depends on biochemical details. Consequently, a variety of patterns of inheritance are observed. The investigation of these patterns and the interplay between genotype and phenotype was pioneered in the nineteenth century by **Gregor Mendel**.

Mendel examined seven traits in peas, each of which was carried on a different type of chromosome. Let's consider a simple **Mendelian cross**. A white-flowered pea from a true breeding line is crossed with a purple-flowered pea from another true breeding line. **True breeding** means the phenotype is constant so long as members of the line are crossed with other members of the line. In the cross between white- and purple-flowered lines of peas, all the progeny are purple. The progeny are called the first filial, or F_1, generation. If the F_1's are crossed among themselves, purple times purple, the second filial, or F_2, generation is 3/4 purple-flowered and 1/4 white-flowered.

The simplest explanation for this **phenotypic ratio** is that a single locus determines the difference in flower color. Each true breeding line would have to be homozygous for a different allele. Let's say that allele **P** causes purple. The true breeding purple line would have the genotype **PP** at the locus for flower color. The true breeding line with white flowers is homozygous for an alternative allele, say **p**, giving the genotype **pp**. Because each parent contributes one copy of each locus to their mutual offspring, all the F_1 would have the same genotype, **Pp**. Clearly the allele for purple color is dominant to the allele for white color because the heterozygote has the purple flower phenotype.

MONOHYBRID CROSS

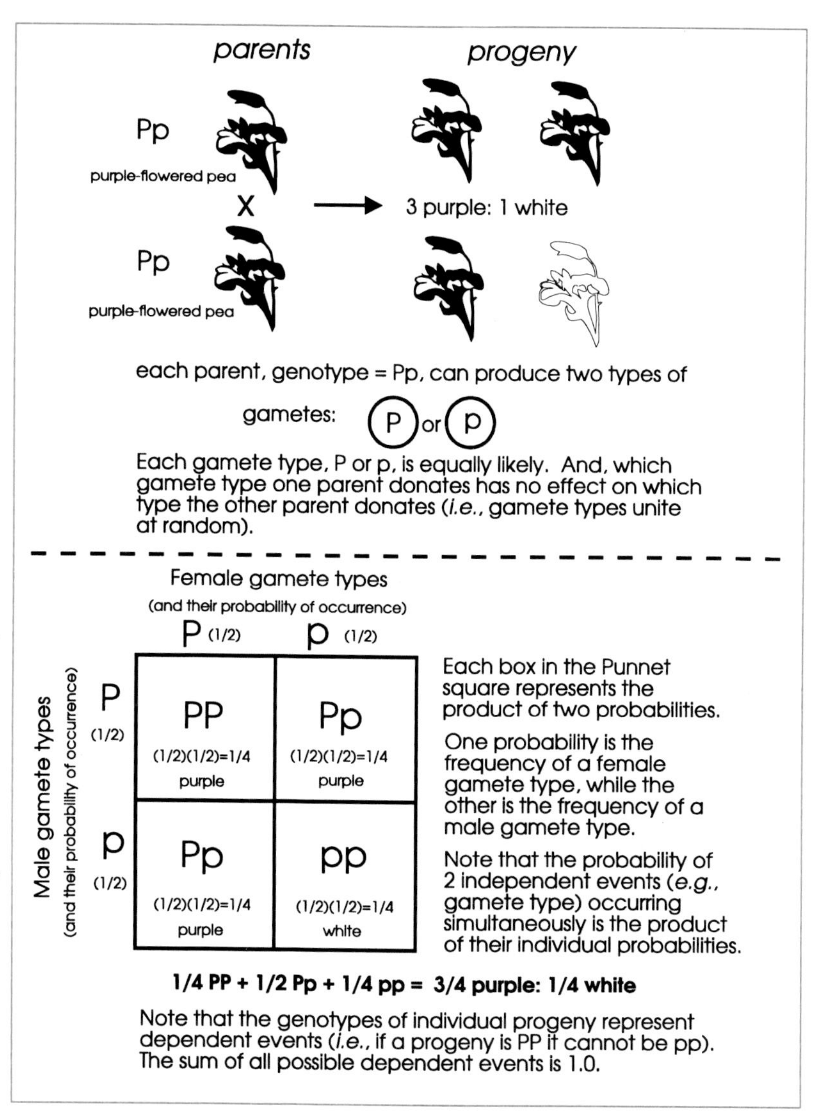

When the F_1s are crossed among themselves this is called an F_1 intercross. Here, the F_1 intercross is between heterozygotes with the genotype **Pp**. To understand how the F_2 generation consists of both white and purple-flowered peas we can construct a **Punnet square**. Simply draw a square divided in half, both vertically and horizontally, so that it has two rows and two columns. At the top of the square, place above one column a **P** and above the other column a **p**. These represent the genotypes of the different female gametes that are possible from a heterozygote. Because of meiosis, half the eggs will carry a **P** allele and half will carry a **p** allele. Now, along the side of the square, place a **P** beside one row and a **p** beside the other row. These represent the genotypes of the male gametes. Again, because of meiosis, half the pollen will carry a **P** allele and half will carry a **p** allele. Now, simply fill in each compartment with one allele from the appropriate column and one allele from the appropriate row. Each compartment represents a possible diploid genotype from the parents crossed.

One compartment has the alleles **PP**, two have **Pp**, and one has **pp**. Three of the four compartments contain genotypes with at least one **P**, meaning that purple flower color would result. Only 1/4 of the progeny have white flowers because only one of the four compartments has the genotype **pp**.

A cross between two heterozygotes at a single locus is called a **monohybrid cross**. The **genotypic ratio** is 1:2:1, as in the example above, one **PP** to two **Pp** to one **pp**. If a trait is determined by a single locus, with two alleles, and with one allele **dominant** to the other, the phenotypic ratio of a monohybrid cross is expected to be 3:1.

If, however, one parent is not a heterozygote, then this 3:1 phenotypic ratio would not be expected. Consider a purple-flowered pea, with the genotype **Pp,** crossed with a white-flowered pea, **pp**. Using a Punnet square we can expect half the progeny to have purple flowers and half to have white flowers.

Let's take a look at another pattern of inheritance, **incomplete dominance**. If a red-flowered carnation is crossed with a white-flowered carnation, all the progeny will have pink flowers. If two pink-flowered carnations are crossed, 1/4 of the progeny will be red, 2/4 pink, and 1/4 white. The phenotype of the heterozygote is intermediate between each homozygote. Let's call the allele for red flower color **R** and the allele for white flower color **r**. When two pink carnations are crossed, we have **Rr** x **Rr**. Filling in a Punnet square gives 1/4 **RR**, 2/4 **Rr**, and 1/4 **rr**. The genotypic ratio, 1:2:1, now matches the phenotypic ratio, one red to two pink to one white.

Incomplete dominance and dominance may result because one allele encodes a functioning enzyme product while the other does not. In the case of purple flower color in peas, enough active enzyme is produced to catalyze the production of a purple pigment whether there is one or two copies of the **P** allele. However, in **Rr** carnations, the single **R** allele cannot give rise to sufficient protein product to produce enough pigment for red flower color, and pink results. Neither the **p** nor **r** alleles produce a protein product that catalyzes pigment production. Therefore, white color results in homozygotes of either allele.

DOMINANCE & INCOMPLETE DOMINANCE

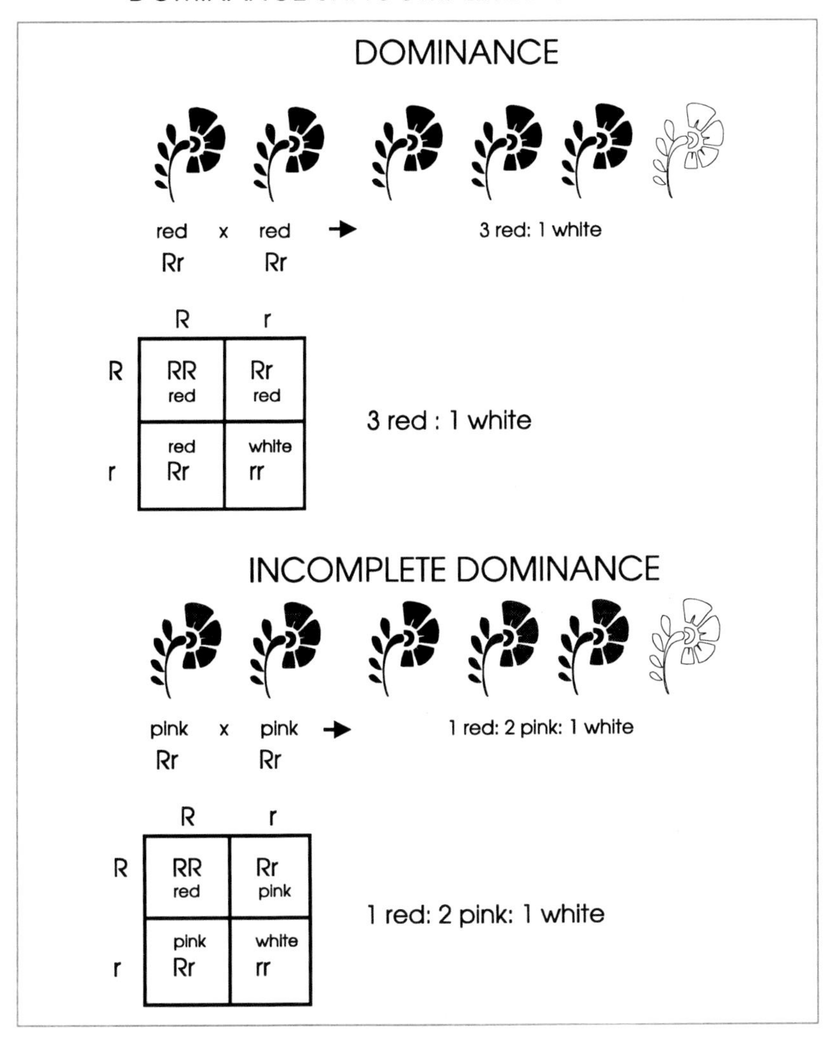

Another pattern of inheritance that mimics incomplete dominance is **codominance**. Here, each allele encodes a fully functional protein product. In heterozygotes, each allele is expressed some or all of the time giving rise to a phenotype intermediate to each homozygote. Consider, for example, coat color in cattle. Red color is produced by the allele $\mathbf{R_1}$ and white by the allele $\mathbf{R_2}$. When a red bull is crossed with a white cow, the progeny are roan. A roan coat has a mixture of red- and white-shafted hairs. Some hair cells express, or transcribe, one allele while other hair cells express the other allele.

Let's consider another example of codominance, **blood type** in humans. Blood type commonly refers to a complex polysaccharide antigen on the surface of blood cells. Individuals with blood type A possess the A antigen whereas individuals with blood type B possess the B antigen. The child of a cross between a blood type A parent and a blood type B parent may have the blood type AB with both antigens present on the surface of blood cells.

Three common alleles at the locus determine blood type in the A/B system. They are designated by the letter **I** which is then superscripted with an **A**, **B**, or **O**. The allele $\mathbf{I^A}$ encodes a protein that gives rise to the A antigen, allele $\mathbf{I^B}$ encodes a protein that gives rise to the B antigen, and allele $\mathbf{I^O}$ gives rise to no antigen. Allele $\mathbf{I^O}$ is recessive to either of the other alleles; that is, $\mathbf{I^A}$ and $\mathbf{I^B}$ are dominant to $\mathbf{I^O}$.

Suppose a man with the genotype, $\mathbf{I^AI^O}$, and a woman with the genotype, $\mathbf{I^BI^O}$, have a child. What's the probability that the child will have a blood type that matches one parent or the other? The father has blood type A because his $\mathbf{I^A}$ allele is dominant to his $\mathbf{I^O}$ allele. The mother has blood type B because her $\mathbf{I^B}$ allele is dominant to her $\mathbf{I^O}$ allele. Therefore, we must consider the possibility that the child will have either blood type A or blood type B. Filling in a Punnet square for this cross gives 1/4 $\mathbf{I^AI^B}$, 1/4 $\mathbf{I^AI^O}$, 1/4 $\mathbf{I^BI^O}$, and 1/4 $\mathbf{I^OI^O}$. The probability that the child has blood type A is 1/4 as is the probability that the child has blood type B. Consequently, the probability that the child has either blood type A or blood type B is 1/4 + 1/4 or 2/4. Each compartment within a Punnet square represents the probability that a certain outcome, here a blood type, will be observed from a particular cross.

In the above cross, it's also possible that the child could have blood type AB, corresponding to the genotype $\mathbf{I^AI^B}$, or blood type O, corresponding to the genotype, $\mathbf{I^OI^O}$.

Now, we see that a Punnet square gives us the expected proportions of genotypes among the progeny of a cross and that this is equivalent to giving us the probability that any given progeny is of a certain genotype. Both proportions and probabilities can only range from 0 to 1, or from 0 to 100 percent. Simply multiply proportions or probabilities by 100 to convert them into percentages. Because a Punnet square lists all the possible combinations of alleles from a cross, the proportions associated with its compartments sum to 1.

BLOOD TYPE PROBLEM

A man of blood type AB and a woman of blood type B want to have a child. If the woman's father had blood type O, what is the probability that their child will be a boy with blood type B?

First, determine the genotype of each parent.

There is only one way to have blood type AB, so the father's genotype mus be: $I^A I^B$.

There are two ways to be blood type B: $I^B I^B$ or $I^B I^O$. However, we can determine the woman's genotype from consideration of her father.

The woman's father had blood type O, which is the recessive phenotype. His genotype had to be:

$I^O I^O$.

Consequently, he had to contribute an I^O allele to his daughter. Therefore, her genotype must be:

$I^B I^O$.

Now, make a Punnet square to illustrate the genotypes and their associated probabilities resulting from a cross of the two parents.

	I^B	I^O
I^A	$I^A I^B$ type AB	$I^A I^O$ type A
I^B	$I^B I^B$ type B	$I^B I^O$ type B

The Punnet square reveals that the probability of a child with blood type B is 2/4.

The probability that a child is a boy is 1/2.

Gender of children and blood type are independent. Therefore, the probability of gender/blood type combinations is the product of their individual probabilities.

Tne answer is: (probability of blood type B) x (probability of a boy) =

2/4 x 1/2 = 2/8 (or **0.25**)

For instance, consider the monohybrid cross, **Aa** x **Aa**. What proportion of the progeny are expected to be of the genotype **Aa**? Filling in a Punnet square gives the genotypic ratio, 1/4 **AA**, 2/4 **Aa**, and 1/4 **aa**. The sum of these three proportions is 4/4 or 1. And, 2/4 or 50% is the proportion that are expected to be of the genotype **Aa**. What is the probability that any given progeny will be of the genotype **Aa**? From the Punnet square we find that 1/2 of the compartments contain this genotype, so 1/2 is the probability of this genotype for any given offspring.

Let's reconsider the blood type example above. A man of type A and genotype $\mathbf{I^A I^O}$, has a child with a woman of type B and genotype $\mathbf{I^B I^O}$. The phenotypic ratio from the Punnet square is 1/4 type AB, 1/4 type A, 1/4 type B, and 1/4 type O, or 1:1:1:1. We found that the probability that the child would have blood type A or B was 1/2 or 1/4 + 1/4. Why did we add the two probabilities associated with blood types A and B?

A child can have only one blood type, if it has type A it cannot have type B. In situations where one outcome precludes others, the different outcomes are said to be dependent. This means that the probability of one outcome, say blood type A, depends on the probabilities associated with other possible outcomes because the sum of the probabilities associated with all possible outcomes must total 1, or 100%, no more, no less. If the probability of one outcome is 95%, then all other dependent outcomes have a (combined) probability of only 5%. If we want to calculate the probability that any one out of a set of dependent outcomes will occur, we add together the probabilities associated with the individual outcomes. In the blood type example above the phenotypic ratio from the Punnet square is one type AB: one type A: one type B: one type O. All four types are equally likely. The probability that a child has blood type AB is 1/4. The probability that a child has blood type AB or A is 1/4 + 1/4 or 1/2. The probability that a child does not have blood type O is 1 - 1/4 or 3/4. Here, we subtracted the probability of blood type O away from one, which is just the sum of the probabilities for all blood types. We could just as well have figured the probability that a child does not have blood type O by summing together the individual probabilities for the other blood types: 1/4 for type AB + 1/4 for type A + 1/4 for type B. Keep in mind that the probabilities associated with genotypes and phenotypes vary with the genotypes of the parents involved in the cross. Therefore, it is wise to figure probabilities by consulting a Punnet square filled out for the specific cross in question.

Some alleles, when present in the homozygous state, are lethal (fatal). The presence of **lethal alleles** is easily detected by phenotypic ratios that occur in thirds or parts of three as opposed to their usual occurence in fourths or parts of four. This is because the compartment of a Punnet square that contains two copies of the lethal allele is ignored when figuring phenotypic ratios. For example, in mice, the allele **C** causes dark color in homozygotes while allele **c** is fatal in homozygotes. Heterozygotes, **Cc**, have yellow coat color. Suppose two yellow mice are crossed. What is the phenotypic ratio expected among the offspring?

Here, each parent is heterozygous, giving a 1:2:1 genotypic ratio in the Punnet square. That is, 1/4 **CC**, 2/4 **Cc**, and 1/4 **cc**. To determine the phenotypic ratio, ignore the **cc** genotypes because they do not survive. Now there are only three compartments in the Punnet square: one with the **CC** genotype and two with the **Cc** genotype. The phenotypic ratio is, then, 1/3 dark coat color and 2/3 yellow coat color.

Many recessive or incompletely dominant alleles are deleterious, or even lethal, in the homozygous state. Many maladies in humans that result from deleterious alleles are said to be inborn errors of metabolism, suggesting the failure of homozygotes to encode functional forms of essential metabolic enzymes. Crosses between close relatives, called **inbreeding**, increase the chance of deleterious alleles occurring in the homozygous state. This is because each parent can acquire, from their common ancestor, copies of the same rare deleterious alleles and pass them on to their (shared) offspring. This is why **hemophilia**, the failure of blood to clot and, thereby, close wounds, became so common among the royal families of Europe.

Sex chromosomes carry genes just as autosomal chromosomes do. Genes on sex chromosomes are called **sex-linked** genes. The very small **Y** chromosome carries very few genes compared with the larger **X** chromosome. There are a large number of **X**-linked genes. For example, it's an **X**-linked gene that provides for the ability of blood to clot at the site of scratches and wounds. A recessive allele at this locus can cause the serious condition of hemophilia, wherein the blood cannot clot. Even minor cuts can lead to life-threatening blood loss by the hemophiliac.

To symbolize **X**-linked alleles we superscript the letter X. Thus, at the locus responsible for hemophilia, the dominant allele is $\mathbf{X}^{H}$ while the recessive allele is $\mathbf{X}^{h}$. Because females have two **X** chromosomes, hemophilia in females only occurs when the genotype is homozygous for the allele $\mathbf{X}^h$. Female hemophiliacs do not survive menstruation. Unlike females, males have only a single **X** chromosome which they receive from their mother. A male hemophiliac has the genotype, $\mathbf{YX}^h$.

INBREEDING

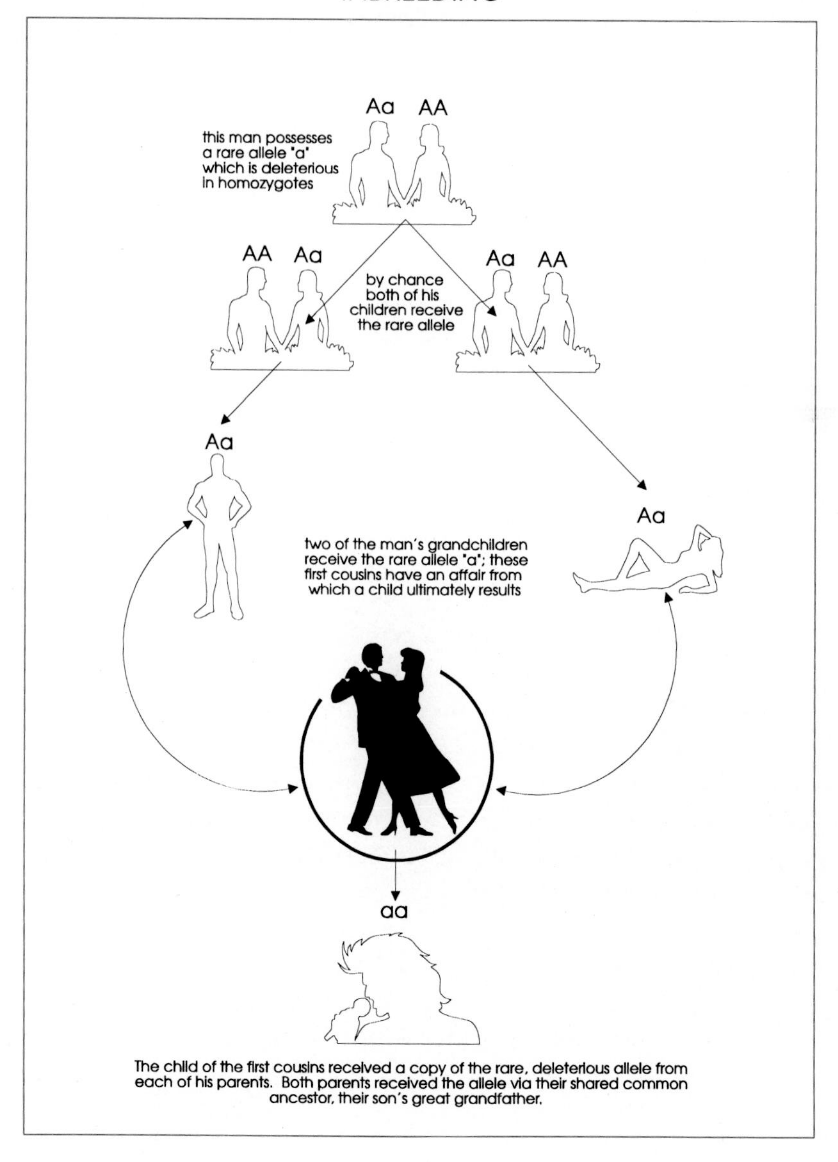

The child of the first cousins received a copy of the rare, deleterious allele from each of his parents. Both parents received the allele via their shared common ancestor, their son's great grandfather.

Suppose a normal woman who is a carrier of the allele $\mathbf{X}^h$ and a normal man have a child. What is the probability that the child will be a hemophiliac? This, and other sex linkage problems, can be solved with a Punnet square. Because the woman is normal in phenotype, meaning her blood clots, but is also a carrier of the allele for hemophilia, her genotype is $\mathbf{X}^H\mathbf{X}^h$. The male is normal so his phenotype must be $\mathbf{YX}^H$. To fill in the Punnet square, use $\mathbf{X}^H$ and $\mathbf{X}^h$ for the female gamete genotypes and use $\mathbf{X}^H$ and **Y** for the male gamete genotypes.

Half of the compartments in the Punnet square contain a **Y** chromosome and correspond to male children. One of these, $\mathbf{YX}^h$, represents a son with hemophilia. The other compartment with a **Y** chromosome corresponds to a normal son with the genotype $\mathbf{YX}^H$. Half of the compartments contain two **X** chromosomes and correspond to daughters. For the cross under consideration, all daughters would be normal with respect to blood clotting ability. A daughter born to this couple would have either an $\mathbf{X}^H\mathbf{X}^H$ or an $\mathbf{X}^H\mathbf{X}^h$ genotype.

Note that different phenotypes among sons and daughters suggest sex linked inheritance. Here, half the sons are expected to be hemophiliacs while none of the daughters would be so afflicted. With **X** linkage, males are more likely to possess traits caused by rare recessive alleles because a male has only one **X** chromosome so his phenotype is determined by the allele on that chromosome. For a female to exhibit a rare recessive trait, she would have to possess two copies of the rare allele, one on each of her two **X** chromosomes.

Now we are ready to answer the original question: what's the probability that a normal male and a normal female, who is a carrier, will have a single child who is a hemophiliac? From the Punnet square, we see that the answer is 1/4, which corresponds to the probability of having an afflicted son. Suppose we ask: what's the probability that their first child is a son? From the Punnet square we see the answer is 2/4 because half of the four compartments contain a **Y** chromosome. Then, suppose we ask: what's the probability that a son will be a hemophiliac? From the Punnet square we see the answer is 1/2 because of the two compartments with a **Y** chromosome, corresponding to sons, only one has the allele $\mathbf{X}^h$.

Now, if we ask what's the probability the first child will be a son and a hemophiliac, we must consider both the child's sex and phenotype with regard to blood clotting ability. The answer is (2/4)x(1/2), giving 2/8 or 1/4.

The probability of being a male by virtue of receiving a **Y** chromosome from the father is 2/4, which is multiplied by the probability of receiving an **X** chromosome from the mother that bears the allele $\mathbf{X}^h$, causing hemophilia. Which chromosome the father contributes to the child, an **X** or a **Y**, is independent of which kind of **X** chromosome the mother contributes. These outcomes are independent because gametogenesis in one parent's body has absolutely no impact on gametogenesis in the other parent's body. To figure the probability that two independent events will occur together, multiply their individual probabilities. This rule, called the **product rule of probability**, is very useful in genetics.

Now let's **review the key words**: genotype, phenotype, homozygote, and heterozygote. Genotype is the combination of alleles present at one or more loci. At any given locus, diploid genotypes can be homozygous, meaning there are two copies of the same allele, or heterozygous, meaning the two alleles differ. Phenotype is an expression of the genotype. The genotype could be thought of as a set of blueprints while the phenotype is what results when those blueprints are read and acted upon. If two individuals have the same genotype their phenotypes may be similar but not necessarily identical because their environments may differ sufficiently to impact the phenotype in different ways. Continuing with our analogy, blueprints may dictate the size and shape of a house but the availability of building materials and paints may dictate what the house actually looks like. The phenotype may or may not be a part of the body of the organism. For instance, different kinds of birds produce different kinds of nests. The size and shape of a nest reflects a bird's genotype just as much as do the colors of its feathers.

Now let's **review the key words**: dominance, recessiveness, codominance, and incomplete dominance. Dominance occurs when the phenotype is the same whether there are one or two copies of a particular allele. If the genotype **Aa** gives the same phenotype as the genotype **AA**, then the **A** allele is dominant and the **a** allele is recessive. With dominance, a monohybrid cross, **Aa** x **Aa**, gives a 3:1 phenotypic ratio. With incomplete dominance and codominace the heterozygote does not resemble either homozygote; rather, it is either intermediate or a mixture of the phenotypes of the homozygotes. In a monohybrid cross, incomplete dominance and codominance produce a phenotypic ratio that matches the genotypic ratio of 1:2:1.

Now let's **review** sex linkage. Sex linkage refers to the presence of loci on one of the sex chromosomes. In humans, most sex-linked genes are, in fact, **X**-linked because the **X** chromosome is much larger than the tiny **Y** chromosome. Sex linkage is revealed in crosses by different phenotypic ratios among male and female progeny. Recessive **X**-linked traits are more common in males than females because females have twice the number of **X** chromosomes giving them twice as many chances of having an **X** chromosome that bears the dominant allele.

CHAPTER 4

MULTI-LOCUS INHERITANCE

Now let's explore the simultaneous inheritance of traits at two loci. **Key words** for this section are: dihybrid cross, epistasis, and pleiotropy. We'll assume the loci are on different chromosomes. That is, they're unlinked. Consequently, we can treat inheritance at the two loci as independent outcomes because of the way meiosis mixes up nonhomologous chromosomes received from the mother and father. For example, consider two loci, the **A** locus with alleles **A** and **a** and the **B** locus, with alleles **B** and **b**. The outcomes at each locus are independent if the genotype at the **A** locus has no influence on the genotype at the **B** locus, and vice versa.

A **dihybrid cross** occurs between two double heterozygotes. For instance, a cross between two peas could be represented by **PpRr** x **PpRr**. The **P** locus determines flower color while the **R** locus determines seed shape. The allele **P**, which causes purple flower color, is dominant to the allele **p**, which causes white flower color. The allele **R**, which causes round seed shape, is dominant to the allele **r**, which causes wrinkled seeds. Dihybrid crosses give a very characteristic phenotypic ratio.

An easy way to determine the ratio is to break the cross down into two smaller monohybrid crosses. For example, **Pp** x **Pp** gives 3/4 purple and

1/4 white flowers. Also, the cross **Rr** x **Rr** gives 3/4 round seeds and 1/4 wrinkled seeds. The result of the dihybrid cross is simply the product of the results for each monohybrid cross: (3/4 purple + 1/4 white) x (3/4 round + 1/4 wrinkled). This gives 9/16 purple/round, 3/16 purple/wrinkled, 3/16 white/round, and 1/16 white/wrinkled. The phenotypic ratio is 9:3:3:1.

DIHYBRID & OTHER MULTILOCUS PROBLEMS

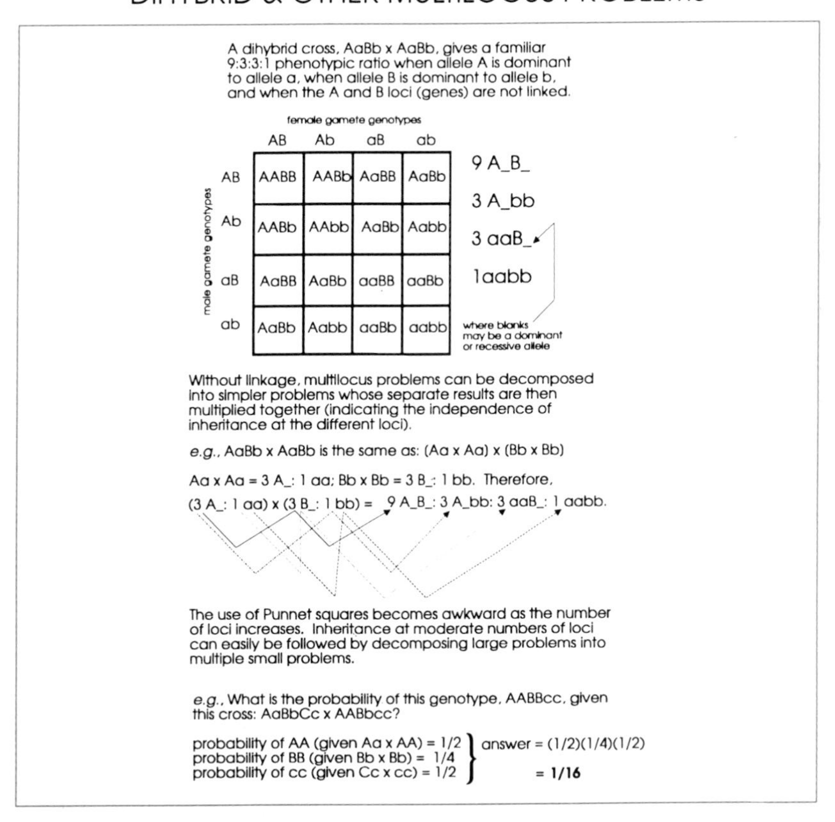

Since each dominant trait occurs at a frequency of 3/4, the probability that an individual possesses the dominant trait at both loci is just (3/4) x (3/4) or 9/16. This was calculated using the product rule of probability. Similarly, the probability that an individual possesses the recessive trait at a locus is 1/4, so the probability that the individual has the recessive trait at both loci is (1/4) x (1/4) or 1/16. The approach of breaking down a large cross into single locus components and then multiplying the results together can be used for any kind of cross in which the loci are not linked.

For instance, suppose **AaBbCc** is crossed with **aabbcc**. This cross involves three loci, the **A** locus, the **B** locus, and the **C** locus. What proportion of progeny are heterozygous at all three loci? Using Punnet squares reveals that for each locus the probability of heterozygosity is 1/2. Therefore, the probability of being heterozygous at all three loci simultaneously is (1/2) x (1/2) x (1/2) or 1/8.

EPISTASIS WITH TWO LOCI

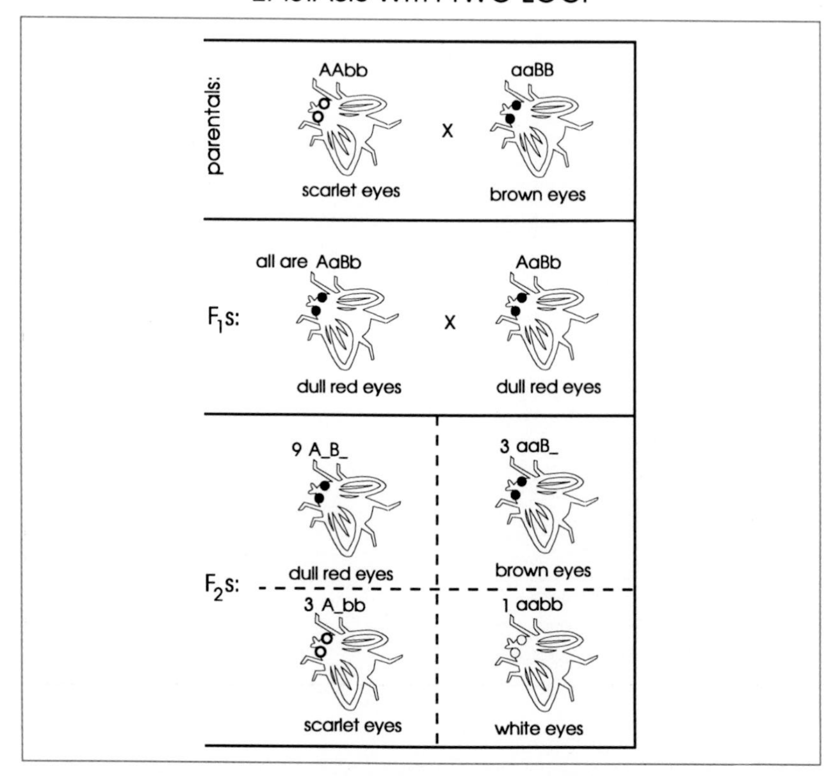

The inheritance of some traits is determined by more than one locus. This condition is called **epistasis**. If two loci are involved, epistasis is revealed in the 9:3:3:1 phenotypic ratio for crosses between double heterozygotes. For instance, in fruit flies the eye color, dull red, requires a dominant allele at each of two loci. Scarlet eye color or brown eye color result if a dominant allele is present at only one or the other locus while white eye color results if an individual has no dominant alleles at either locus. The cross dull red times dull red yields the phenotypic ratio 9 dull red: 3 scarlet: 3 brown: 1 white if each dull red parent was a double heterozygote. Some **epistatic interactions** give phenotypic ratios such as

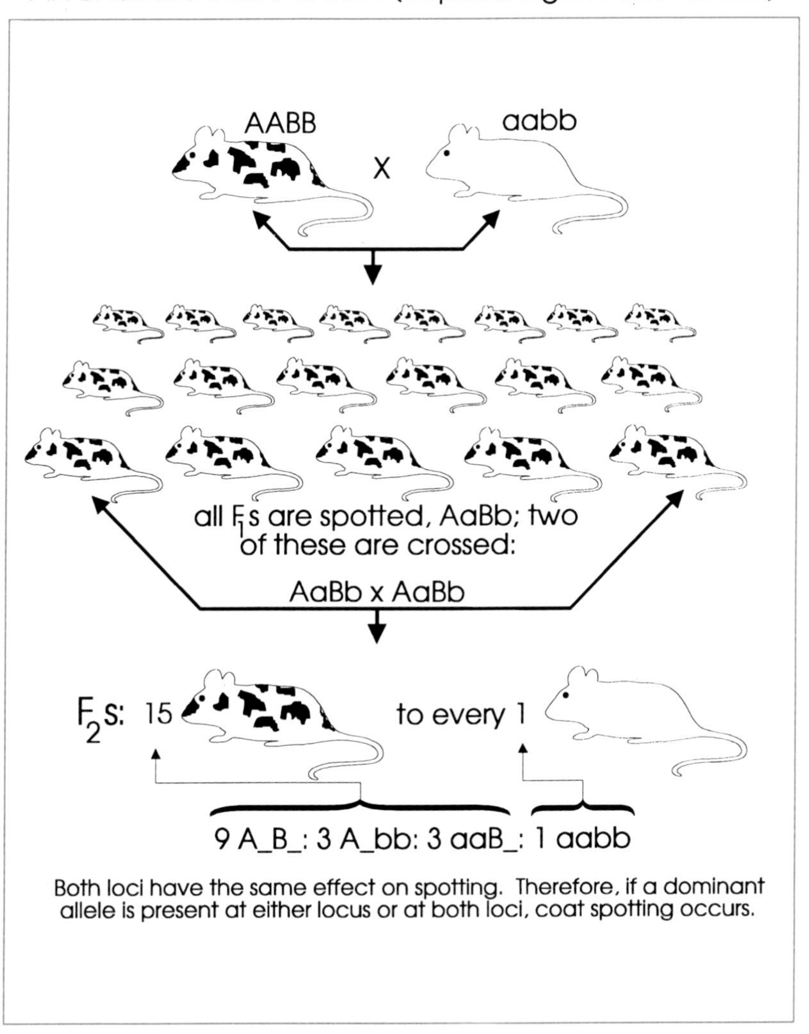

9:7 or 15:1. Phenotypic ratios that come in parts of 16 are indicative of two loci.

Interaction between loci to determine a trait is very common. In fact, dozens of loci may interact to determine some phenotypic traits. On the other hand, a single gene can affect multiple traits. This is **pleiotropy**.

Now let's **review the key words**: dihybrid cross, epistasis, and pleiotropy. A dihybrid cross is a cross between two double heterozygotes. A double heterozygote is an individual that is heterozygous at both of the loci being followed in the cross. Dihybrid crosses yield a characteristic 9:3:3:1 phenotypic ratio if inheritance at each locus is characterized by dominance and not incomplete dominance nor codominance. The 9 refers to genotypes with at least one dominant allele at each locus, the 3s refer to genotypes with at least one dominant allele at one locus and two recessive alleles at the other locus, while the 1 refers to a genotype that is homozygous for the recessive alleles at both loci. Epistasis is the condition in which phenotypic differences, at a single trait, are determined by allelic differences at more that one locus. Pleiotropy is the contribution of a single locus to multiple traits.

CHAPTER 5

POPULATION GENETICS

Sometimes it's necessary to consider the array of genotypes across a whole population rather than just within the progeny of a single cross. Following genotypes at a locus within a sexually reproducing population is our next topic. **Key words** are: population, gene pool, allele frequency, genotype frequency, and Hardy-Weinberg equilibrium.

A **population** is a group of potentially interbreeding individuals. Interbreeding causes the sharing and mixing of alleles between families and between locales. This sharing and mixing results in genetic continuity, meaning alleles common in one locale are likely to be common in another. All the alleles, of all the genes, in the various locales united by interbreeding constitute the **gene pool**. Each population has its own definable gene pool. The spatial scale across which interbreeding maintains genetic continuity is called **population structure**. Organisms with limited dispersal abilities and organisms occupying fragmented habitats tend to have their populations structured on very small geographic scales while mobile organisms and those occupying continuous habitats have populations structured on large spatial scales.

Gene pools are characterized by the **frequencies of alleles** at loci. Let's imagine a locus with two alleles, **A**, which is dominant, and **a**, which is recessive. The frequency of an allele is a measure of its average representation among the genotypes of the population. The frequency of allele **A** is the quotient, the number of **A** alleles among all genotypes in the sample divided by twice the number of individuals in the sample. The sample may be the whole population or a representative subset. Division is by twice the number of individuals since each diploid genotype has two copies of each locus. Similarly, the frequency of allele **a** is the number of **a** alleles among all genotypes in the sample divided by twice the number of individuals in the sample. Allele frequencies can only range from 0 to 1, therefore, the frequencies of the two alleles will always sum to 1. Usually, the frequency of the dominant allele is symbolized by p while the frequency of the recessive allele is symbolized by q. Therefore, $p + q = 1.0$. If the frequency of one allele is known, the other can be determined by subtracting the known value from 1.

Genotypes also have frequencies that range from 0 to 1. The frequency of any genotype, such as **AA**, is the quotient, the number of individuals with the genotype divided by the total number of individuals. There is no need to divide by twice the number of individuals because each individual has a single genotype.

A Punnet square can be used to derive an equation that permits **genotype frequencies** to be calculated from allele frequencies. We'll let p be the frequency of allele **A** and q be the frequency of allele **a**. Next, we'll assume that the frequencies of the alleles are the same in each sex. Now, we can construct a Punnet square with the column and row labels being p and q instead of the allele designations, **A** and **a**. Filling in the four compartments of the square as before, we get one compartment with pp, two with pq, and one with qq.

Each compartment represents the probability that a particular combination of alleles would result. So long as males and females do not mate on the basis of their genotype at the **A** locus, the allele donated by the female parent has no bearing on which allele is donated by the male parent. Therefore, when we multiply allele frequencies from males and females we are determining the frequency or probability of the corresponding genotype. The compartment containing pp, or p^2, provides the frequency of the genotype **AA** in the next generation. Notice there are two compartments containing pq. This is because the genotype **Aa** can be made in two ways:

A from the father and **a** from the mother or **a** from the father and **A** from the mother.

Adding all the compartments from the Punnet square gives us the equation: $p^2 + 2pq + q^2 = 1$. The sum equals one because the equation accounts for the frequencies of all possible genotypes. Or, look at it this way: $p + q = 1$, so the column labels sum to 1 as do the row labels. Consequently, all we've done is multiply 1 times 1, which gives 1. The term p^2 gives the frequency of **AA**, $2pq$ gives the frequency of **Aa**, and q^2 gives the frequency of **aa**.

HARDY-WEINBERG EQUILIBRIUM

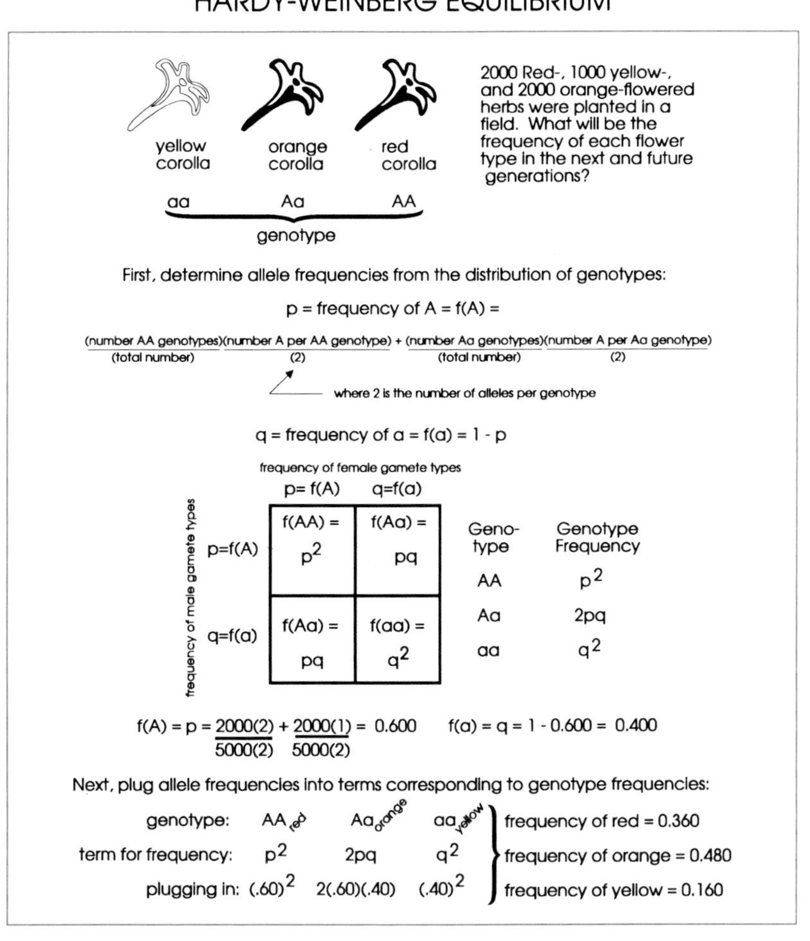

Let's try a problem. Suppose allele **P** is dominant and causes purple flower color while allele **p** is recessive and causes white flower color in **pp** homozygotes. If a random sample of a population yielded 84 purple- and 16 white-flowered plants, what proportion of the plants are heterozygous at the **P** locus? Because dominance causes the heterozygotes to be indistinguishable from **PP** homozygotes, it's necessary to first determine the values for p and q and then plug these values into the term $2pq$ which represents the frequency of heterozygotes.

The frequency of the genotype **pp** is easily determined as 16, the number of white-flowered plants in the sample, divided by 100, the total sample size. This gives 0.16 which is equal to q^2. q, then, is the square root of q^2, which is the square root of 0.16 or 0.4. With q equal to 0.4, p has to be 0.6, which is $1 - q$. Remember $p + q$ equals 1. Now, we can plug these values into the term $2pq$, to determine the frequency or proportion of heterozygotes, which is (2) x (0.6) x (0.4), or 0.48.

As long as the values of p and q do not change under pressure from evolutionary forces, random mating with regard to the locus in question will establish a stable set of genotype frequencies that will not change from generation to generation. The establishment of stable frequencies is called the **Hardy-Weinberg equilibrium**. Equilibrium is a steady state or state of no further change. For autosomal loci, the equilibrium for genotype frequencies is established after only one generation of random mating.

Now let's **review the key words**: population, gene pool, allele frequency, genotype frequency, and Hardy-Weinberg equilibrium.

A population is a group of potentially interbreeding individuals. Through sexual reproduction, they share access to the gene pool, a common set of alleles across all loci. Each allele within the gene pool has a frequency which is the number of that allele, in genotypes of all individuals, divided by twice the number of individuals. Similarly, all genotypes have a frequency which is the number of that genotype divided by the number of individuals. Allele frequencies and genotype frequencies are related by the equation $p^2 + 2pq + q^2 = 1$. This equation defines the Hardy-Weinberg equilibrium which is the stable set of genotype frequencies that occur with stable allele frequencies. Here, p is the frequency of one allele and p^2 is the frequency of the homozygote for that allele. Similarly, q is the frequency of the other allele and q^2 is the frequency of the other homozygote. The remaining term, $2pq$, is the frequency of heterozygotes.

CHAPTER 6

RECOMBINANT DNA TECHNOLOGY

Theoretical and applied advances continue to be made in both population genetics and in its allied field, quantitative genetics. Quantitative genetics is concerned with the inheritance of polygenic traits. Both population genetics and quantitative genetics are especially important with respect to evolutionary theory. However, in no area of genetics has our understanding advanced faster than in molecular genetics, particularly with respect to our ability to manipulate the genetic constitution of organisms. Genetic engineering is the term applied to such manipulation. Next, we discuss a few of the tools and techniques that make this possible. **Key words** for the next section are: restriction enzyme, vector, ligase, transformation, library, cDNA, and PCR.

Restriction endonucleases, or **restriction enzymes**, are proteins that cut through double stranded DNA at sites where specific, short sequences of bases occur. Each kind of restriction enzyme has its own, specific, and often unique, **recognition sequence**. The recognition sequence is usually four to six bases long and is the point at which the enzyme binds to DNA and cuts through the DNA. For example, the enzyme, *Eco*R I, recognizes the six-base sequence, GAATTC. Under proper conditions, this enzyme would cleave through strands of DNA at every point where the sequence GAATTC occurs. With over 100 different restriction enzymes commercially available, specific genes can easily be sliced out of a chromosome by digesting the DNA with appropriate restriction enzymes whose recognition sequences lie outside of, but very near, the genes of interest.

Restriction enzymes occur naturally in **bacteria** where they appear to have evolved as a defense against viruses called **bacteriophages**. But, once isolated and purified, a restriction enzyme can cut any kind of DNA, including that of mammals. Most restriction enzymes cleave DNA with **staggered cuts**. To understand a staggered cut imagine that the two strands of DNA are two rows of bricks. Bricks are laid such that one row is staggered by 1/2 brick relative to the row above or below it. Restriction enzymes cut the DNA in a zig-zag-like path as if two rows of bricks were cleaved by only cutting through the mortar.

The reason for this staggered cutting pattern is that most recognition sequences are palindromic, meaning that the sequence on one strand of DNA is the same as the complementary sequence on the other strand if it's read in the opposite direction. For instance, the recognition sequence for the enzyme, *Eco*R I, is GAATTC. Its complement is CTTAAG, which is the same sequence as the recognition sequence, only backwards. The enzyme cuts each strand between the G and the adjacent A. Consequently, on each side of the cleavage site there are four bases of single strand DNA dangling off the end of what is otherwise double stranded DNA. The four bases on one end are complementary to those on the other end. Because these two ends could hydrogen bond together, they are referred to as "**sticky ends**".

Now imagine two different batches of DNA, each digested with the same restriction enzyme so that their sticky ends are compatible. If the two batches of DNAs are mixed, fragile, hybrid molecules would form due to hydrogen bonding between sticky ends. In this way, DNA from one organism can be inserted into the DNA of another organism. However, covalent linkages still need to be made between the backbones of the foreign DNAs.

The enzyme, **T_4 DNA ligase**, is often used to catalyze covalent bond formation within both strands of a hybrid DNA molecule. Ligase requires the energy of ATP to drive the formation of covalent bonds in the DNA backbone. This ligase is the product of a viral gene and may represent a

DNA-repair defense of the virus against attack by bacterial restriction enzymes.

Together, restriction enzymes and ligase make it possible to almost surgically remove a gene from its flanking DNA and then reinsert it into foreign DNA that's capable of being replicated and expressed within a living cell. Often, genes of interest are inserted into a carrier DNA called a **vector**. The most common vectors are bacteriophage DNA and plasmids. A **plasmid** is a small circular DNA of a few thousand base pairs. Vectors are genetically engineered with at least one restriction enzyme recognition sequence to facilitate the insertion of foreign DNA.

RESTRICTION ENZYMES

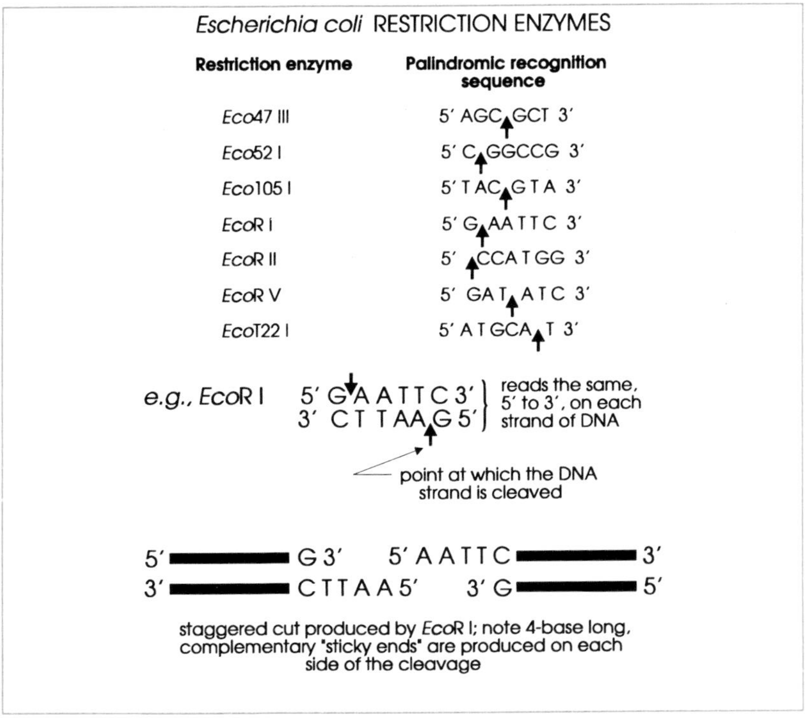

Escherichia coli RESTRICTION ENZYMES

Restriction enzyme	Palindromic recognition sequence
*Eco*47 III	5′ AGC↑GCT 3′
*Eco*52 I	5′ C↑GGCCG 3′
*Eco*105 I	5′ TAC↑GTA 3′
*Eco*R I	5′ G↑AATTC 3′
*Eco*R II	5′ ↑CCATGG 3′
*Eco*R V	5′ GAT↑ATC 3′
*Eco*T22 I	5′ ATGCA↑T 3′

e.g., *Eco*R I 5′ G↓AATTC 3′ / 3′ CTTAA↑G 5′ — reads the same, 5′ to 3′, on each strand of DNA

point at which the DNA strand is cleaved

5′ ▬▬▬ G 3′ 5′ AATTC ▬▬▬ 3′
3′ ▬▬▬ CTTAA 5′ 3′ G ▬▬▬ 5′

staggered cut produced by *Eco*R I; note 4-base long, complementary "sticky ends" are produced on each side of the cleavage

Vectors are necessary for the **transformation** of an organism with foreign DNA. A cell is transformed when it takes up foreign DNA which is then either replicated along with the cell's chromosomes or integrated into the cell's chromosomes. Plasmid vectors usually exist autonomously, that is, outside the cell's chromosomes, while viral vectors become integrated into the host chromosome. The process of **integration**, whereby the vector DNA becomes inserted into host DNA, entails recombination.

SUBCLONING INTO A PLASMID

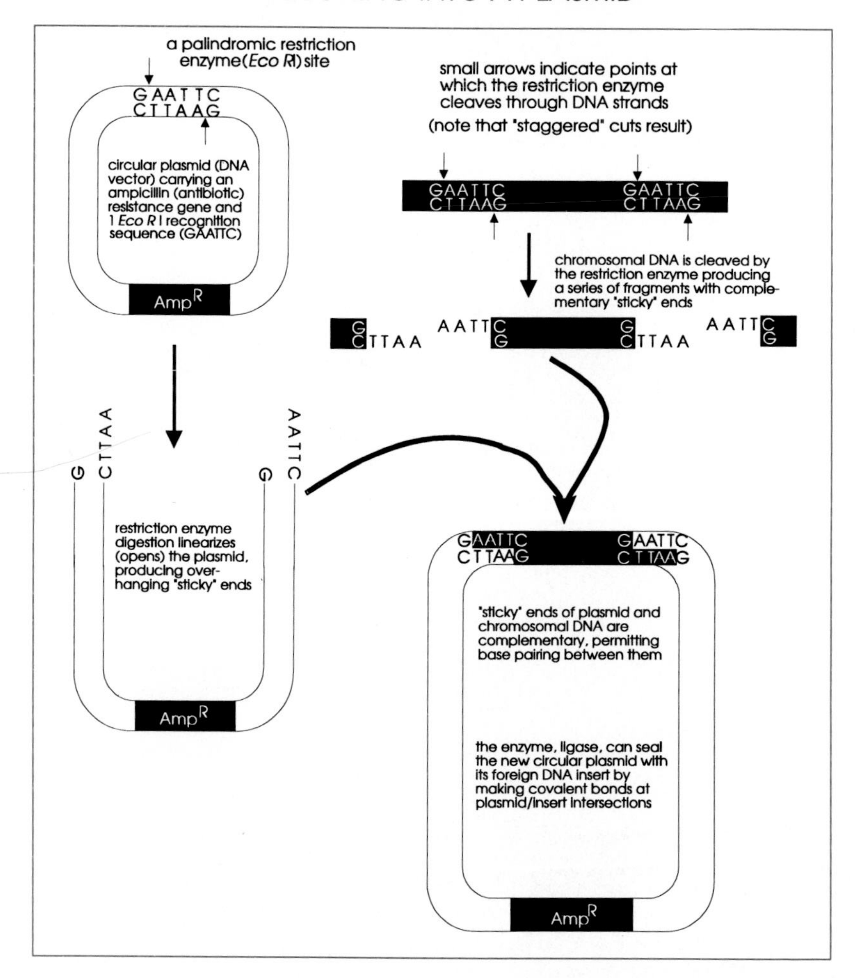

Genetic engineering takes on many forms. Some commercially available drugs, such as the mammalian hormone, insulin, are now produced in bacteria. The gene for insulin is carried on a plasmid vector with which the bacteria have been transformed. The insertion of genes into plasmid vectors is referred to as subcloning. Vertebrates can also be genetically engineered by transforming their egg cells with genes carried on viruses. **Retroviruses** are normally used for this because they naturally integrate into the host chromosomes. Transformation caused by viruses is called **transfection**. It's also possible to achieve low levels of integration into host chromosomes without linking the genes of interest to vectors. For example, DNA can be coated onto the surface of gold pellets and shot into a cell using a gene gun.

LIBRARY CONSTRUCTION

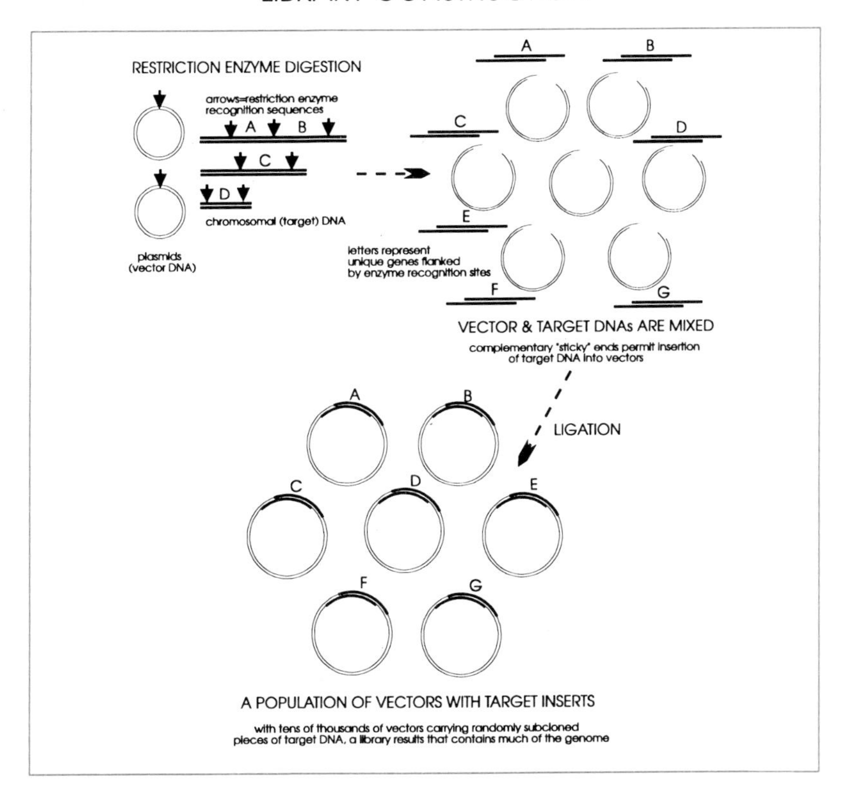

Bacteriophage vectors are commonly used to produce gene libraries. A **library** consists of thousands of vector DNAs. Each vector carries a random piece of the total DNA of the organism of interest. If the library is large enough all of the target organism's genes and other DNA elements will be represented. Bacteriophage DNA is packaged into phage protein coats that can later be used to infect bacteria. This permits the library to be screened for particular genes and allows for the future harvesting of large quantities of those genes for further study.

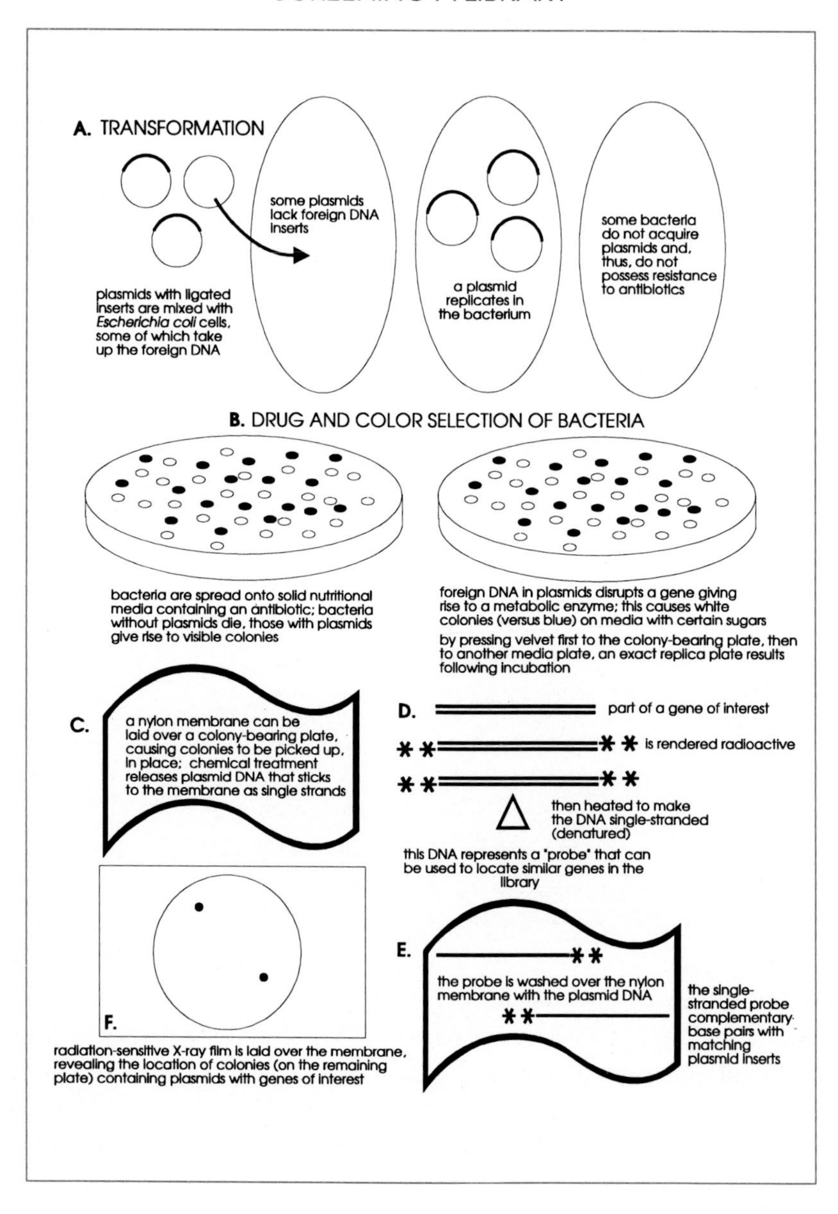

Some libraries, called **cDNA** libraries, are made from messenger RNA transcripts and not from DNA. The transcripts are isolated and then converted into DNA with the retroviral enzyme, **reverse transcriptase**. Reverse transcription is essentially DNA replication using an RNA template as a guide for the production of a second, complementary strand of DNA. Consequently, cDNA libraries are devoid of the nongenic DNA that constitutes a high proportion of the DNA of other libraries. cDNA libraries derived from different tissues or from different developmental stages vary according to which genes are being expressed in given tissues at a given time.

Obtaining specific genes for study or subcloning has been greatly facilitated by the advent of the **polymerase chain reaction**. The polymerase chain reaction, or **PCR**, is a means of synthesizing large quantities of specific target genes or other DNA sequences. PCR uses a heat-insensitive DNA replication enzyme originally acquired from bacteria living in thermal streams.

To start a polymerase chain reaction, a solution of DNA is heated to almost the boiling temperature of water. This breaks the weak hydrogen bonds that hold the two strands of a DNA molecule together. The DNA is now said to be **denatured**. Next, the temperature is greatly reduced, allowing a specific **primer** to bind, or **anneal**, to each strand. A primer is a short, single, strand of DNA built to be complementary to a short stretch of bases on one or the other side of the gene of interest. The DNA replication enzyme, usually *Taq* polymerase, then builds a new complementary strand using the primer as its starting point. Thus, replication procedes through the gene of interest. Replication is once again terminated by raising the temperature of the solution so that the newly synthesized DNA strand and the old strand separate.

If the temperature is again lowered, replication through the gene of interest will again commence. However, this time there are twice as many single-stranded templates as before to which primers can anneal. Each round of denaturation, annealing, and replication doubles the number of DNA templates for the next **cycle** in the chain. After 20 to 40 cycles, almost all of the DNA present in the experimental solution will consist of the gene of interest.

Now let's **review the key words**: restriction enzyme, vector, ligase, transformation, library, cDNA, and PCR. A restriction enzyme catalyzes the cleavage of double stranded DNA at sequence-specific recognition sites. Most restriction enzymes produce staggered cuts with single-stranded loose ends that can base pair with the loose ends of other pieces of DNA cut with the same enzyme. These "sticky ends" permit the insertion of one DNA into another. The enzyme ligase catalyzes the sealing of the new hybrid DNA. This sealing process is called ligation. The ligation of foreign DNAs into a vector DNA makes it possible to transform cells. Vectors are DNA mole-

POLYMERASE CHAIN REACTION

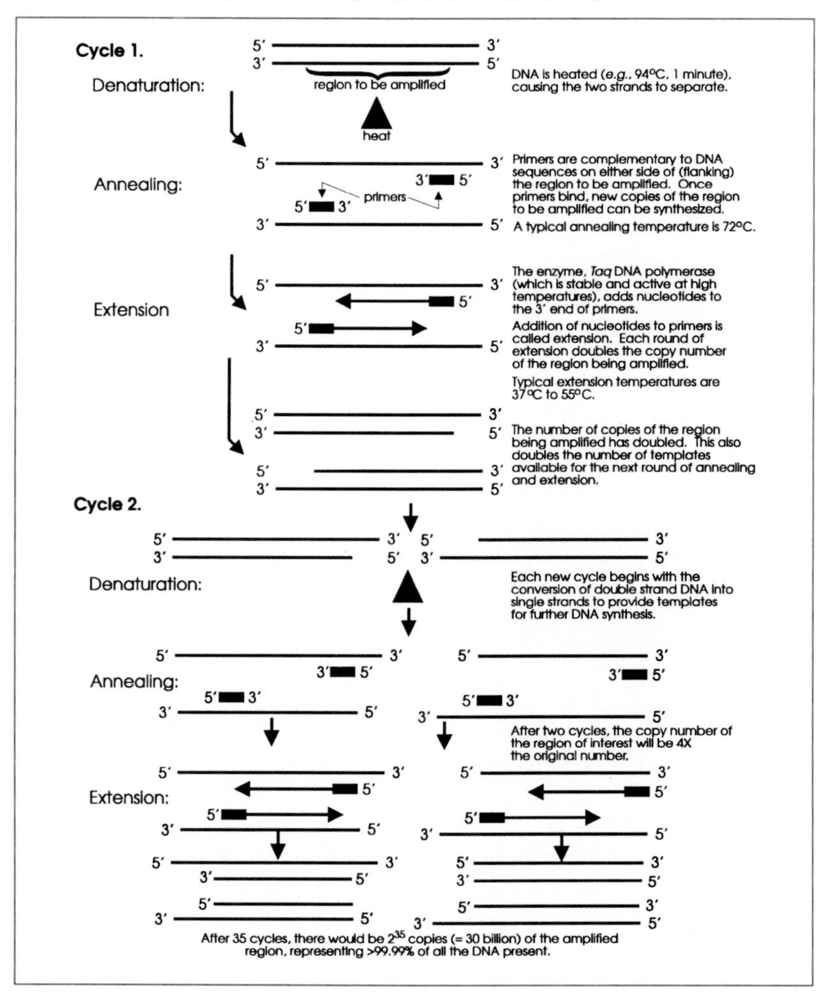

cules that can carry genes from one organism to another. Using vectors or gene guns to experimentally engineer the genetic constitution of an organism is called transformation.

Vectors are also used to produce gene libraries. A library is all or most of the DNA of an organism that has been cut into small pieces with a restriction enzyme, and then ligated into vectors which can be screened to retrieve genes of interest. A cDNA library contains only genes for proteins since it's derived from messenger RNAs that have been converted into DNA by the enzyme, reverse transcriptase.

The polymerase chain reaction, or PCR, is a DNA-replication process using a heat stable form of the enzyme, DNA polymerase. Cycles of heating and cooling permit repeated rounds of replication of the same short sequence of DNA. Each successive cycle doubles the number of target templates for the next round of replication. Thus, large quantities of specific genes can be amplified by using primers that exactly flank the genes of interest. Primers are short, single strands of DNA that base pair to a single-stranded DNA template to initiate DNA replication.

CHAPTER 7

NATURAL SELECTION & THE MODERN SYNTHESIS

Since the beginning of the twentieth century the study of genetics has been united with the study of evolution. This union is called the **modern synthesis**. Primarily, the modern synthesis grew out of the works of three population geneticists, **J. B. S. Haldane**, **Sewell Wright**, and **Ronald A. Fisher**. They described how the frequencies of alleles and genotypes would change under the influence of evolutionary forces. Next, we discuss these evolutionary forces and the **key words** for the next section are: evolution, natural selection, and adaptation.

Evolution entails change. Often evolution is trivially defined as a change in allele frequency within a population, however minor the change. Depending on the evolutionary forces acting, minor changes in allele frequency can be reversed from one generation to the next, or even within a generation, so that the net change is zero across a number of generations - thus, no evolution. Less trivially, **evolution** is defined as any longterm, cumulative genotypic change within a population. Thus, genotypic changes underlie the phenotypic changes that are observed and measured. Evolution always refers to changes within a population of individuals. Single individuals do not evolve, populations do.

The modern synthesis is not just a genetic version of the idea that humans evolved from ape-like ancestors. Nor is it limited to the study of **speciation**, which addresses how a new species originates within a subset of the populations of a pre-existing species. Evolution also refers to the changes in the bill shapes of birds on oceanic islands, the changes in the banding patterns of snails on tropical beaches, the changes in the nectar rewards jungle flowers offer to pollinating bees, and virtually every other trans-generational phenotypic change observed or inferred to have happened. The study of evolution encompasses the origin of cellular life, perhaps from crystalline mineral ancestors, the physiological and ecological adaptations of organisms to their specific environments, and the causes of speciation and extinction. The rate of evolution is slow for most organisms because evolutionary forces are frequently weak and because different forces may cancel each other out. Consequently, evolution may require hundreds or thousands of generations. Because the generation times of many organisms are often a substantial proportion of our own lifetime, it's rare to observe evolution in nature. However, in the laboratory, the strength of evolutionary forces can be manipulated so that the evolution of a phenotype may be observed if experiments focus on organisms with short generation times, such as bacteria or fruit flies.

Often, evolution is studied using a comparative rather than an experimental approach. With the **comparative approach**, the impact of an evolutionary force is determined by evaluating phenotypes across a spectrum of individuals, populations, or species for whom the force is believed to vary. For instance, consider the hypothesis that the loss of flight in island-dwelling insects is an adaptation that prevents them from flying out over waters where they would perish. This could be tested by comparing the number of flightless insect species on islands versus mainlands.

Comparative arguments were used to great, and overwhelming, advantage in the influential works of the 19th century biologist, **Charles Robert Darwin**. Darwin published *On the Origin of Species by Means of Natural Selection*, in 1859, *The Descent of Man and Selection in Relation to Sex,* in 1871, and numerous other books on, among other things, orchids, barnacles, and animal expressions. Darwin is best known for the concept of **natural selection** and shares the credit for its formulation with the naturalist, **Alfred Russell Wallace**.

Darwin observed that many more offspring were produced than could survive on available resources. He was encouraged in these observations by **Thomas R. Malthus**' book, *An Essay on the Principle of Population*, which argued that human populations grow at exponential rates while our ability to harvest resources increases at the slower linear rate. Therefore, Darwin envisioned a struggle for survival. Individuals with traits that provided an edge in the competitive struggle for resources were said to be more **fit** for their particular environment. The more fit individuals would, on average, out reproduce those without the advantageous traits. Therefore, if traits conferring higher **reproductive success** were inherited, the frequency of the advantageous traits would increase from one generation to the next. This is the essence of the concept of natural selection.

NATURAL SELECTION

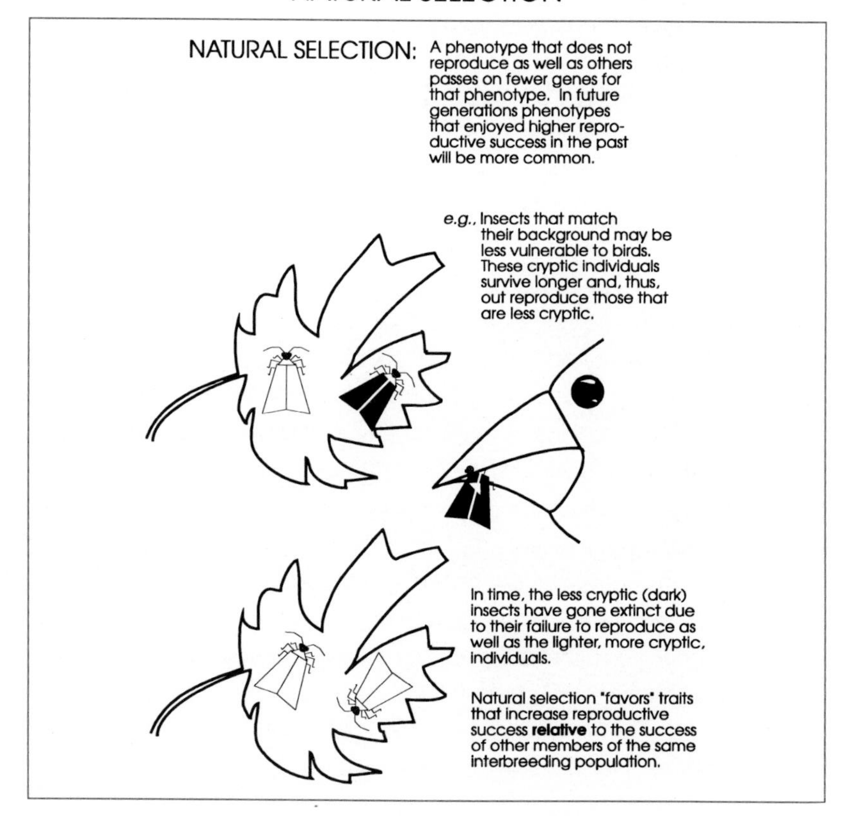

Natural selection is the change in frequency of traits as a consequence of differences in reproductive success conferred by those traits. For natural selection to occur individuals must vary in their reproductive success and at least some of the variation in reproductive success must be due to phenotypic differences. Evolution can then occur if at least some of the phenotypic differences are due to genotypic differences.

For example, individuals within a population of moths might vary with respect to how well their wing coloration matched the bark of trees on which they rested during the day. The more cryptic individuals would be less frequently detected by predators than individuals that did not match their bark background. Thus, the more cryptic an individual's coloration, the greater its chances of surviving to reproduce. If over a number of generations the moth's population was composed of more and more cryptic individuals, this would reflect both the greater reproductive success of cryptic individuals, due to their greater ability to survive, and would also reflect the inheritance of cryptic coloration.

In fact, during the latter half of the 19th century hundreds of species of moths in Europe and the United States evolved darker wing coloration in an apparent response to increasing levels of dark soot deposits on tree bark. The soot came from the explosion of industrial activity, fueled by coal, known as the industrial revolution. The evolutionary response of moth species, such as the British peppered moth is called **industrial melanism**.

Traits that evolve in response to natural selection are called **adaptations**. Natural selection is said to *favor* adaptations that increase survival, fertility (the number of viable gametes produced) and fecundity (the number of offspring produced). Environmental contingencies such as harsh weather, competitor species, predators, evasive actions of prey, *etc.*, are said to *select for* particular adaptations.

The presence of a selection pressure, such as the need to better hide from predators, does not, of itself, guarantee evolution by natural selection. There must be **genetic variation** for the phenotypic traits in question. That is, there must be alternative alleles at the loci determining the traits in question. If there is only one allele per locus, all individuals have the same genotype. In this case, any phenotypic differences that might exist among parents are due to the environment and cannot be passed on to their offspring. R. A. Fisher's *Fundamental Law of Natural Selection* states that the rate at which natural selection promotes evolution is directly proportional to the amount of genetic variation for the traits in question.

Genetic variation arises from mutations. **Mutations** are changes in the DNA code that occur without regard to the needs of the organism. Consequently, most mutations are harmful and natural selection acts to reduce the frequency of the deleterious alleles. However, some mutations are beneficial and provide the opportunity for the evolution of adaptations. Mutations are called the *raw material* of evolution.

Now let's **review the key words**: evolution, natural selection, and adaptation. Evolution is the long term change in the frequency of phenotypes within a population. Evolution is a consequence of changes in allele frequencies that are mediated by so called evolutionary forces. Charles Darwin argued the role of natural selection in evolutionary change. Natural selection is the change in allele frequency caused by genotypic differences in reproductive success. Changes in allele frequency are reflected in new genotypic frequencies which are then reflected in new phenotypic frequencies. Traits that evolve under natural selection pressures are adaptations. Adaptations permit organisms to exploit their environments in ways that are reflected in greater survival, fertility, or fecundity. Adaptive change requires genetic variation that ultimately arises from mutations in the genetic code.

CHAPTER 8

MUTATIONS & GENETIC VARIATION

There are many kinds of mutations. **Key words** for our discussion of mutations are: point mutation, silent mutation, neutral mutation, frameshift, and transposable element. **Point mutations** affect a single base within the DNA. **Transversions** substitute a purine base, adenine or guanine, for a pyrimidine base, cytosine or thymine, or *vice versa*. **Transitions** substitute a purine for a purine or a pyrimidine for a pyrimidine. A transition or transversion may or may not affect the structure of the protein encoded by a structural gene. If the amino acid sequence of the protein is not affected the mutation is said to be **silent**. The mutation is **neutral** if it causes one amino acid to be substituted for another, but without effect on protein function. Natural selection is blind to silent and neutral mutations and, therefore, does not impact the frequency of the new alleles. However, a population may accumulate neutral mutations that, under different conditions, are no longer neutral and become subject to natural selection.

Base substitutions that impact protein function are not neutral and are subject to natural selection. Sometimes a single base substitution can have a major impact on the protein encoded by a gene. For example, a single amino acid change in individuals with sickle cell anemia destroys the conformation, or shape, of hemoglobin that is necessary for oxygen delivery. Sometimes a single base change can destroy important transcriptional signals such as start and stop codons that play critical roles in the proper reading of messenger RNA transcripts during the process of translation.

FRAMESHIFT MUTATION

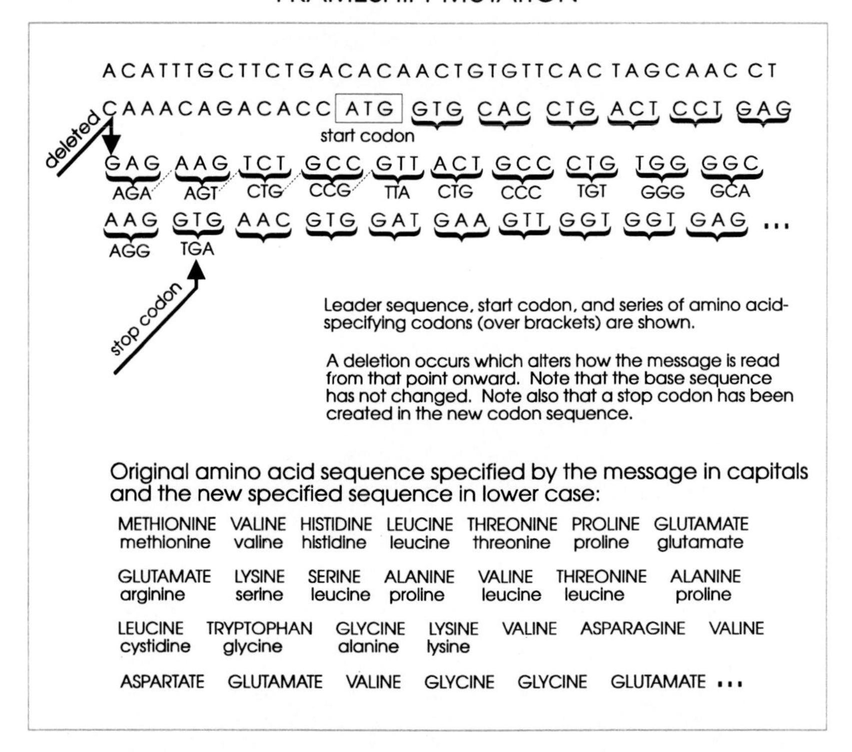

Some point mutations cause the deletion or insertion of one nucleotide in the DNA. This can cause a **frameshift** during translation wherein the sequence of 3-base codons in messenger RNA is misread from the point of the mutation to the end of the transcript. These and other point mutations occur under a number of circumstances such as: when DNA is struck by ionizing radiation from the sun, when chemical mutagens interact with DNA, and when bases are mismatched during DNA replication. Because DNA replication enzymes correct mismatched bases with proofreading functions that detect improperly paired bases, very few point mutations occur during DNA replication.

Mutations also arise in other ways. When a DNA sequence is repeated in tandem, that is, one copy right after another, some of the repeated units can be either duplicated or deleted from a chromosome. Mutations arise from **unequal exchange** which follows the misalignment of homologous chromosomes during recombination or from **replication slippage** which is the looping out of part of one strand of DNA, relative to its complement, during replication. These processes occur at a relatively high frequency and can cause the replacement or turnover of the repetitive sequences that can constitute a large proportion of the genome.

UNEQUAL EXCHANGE

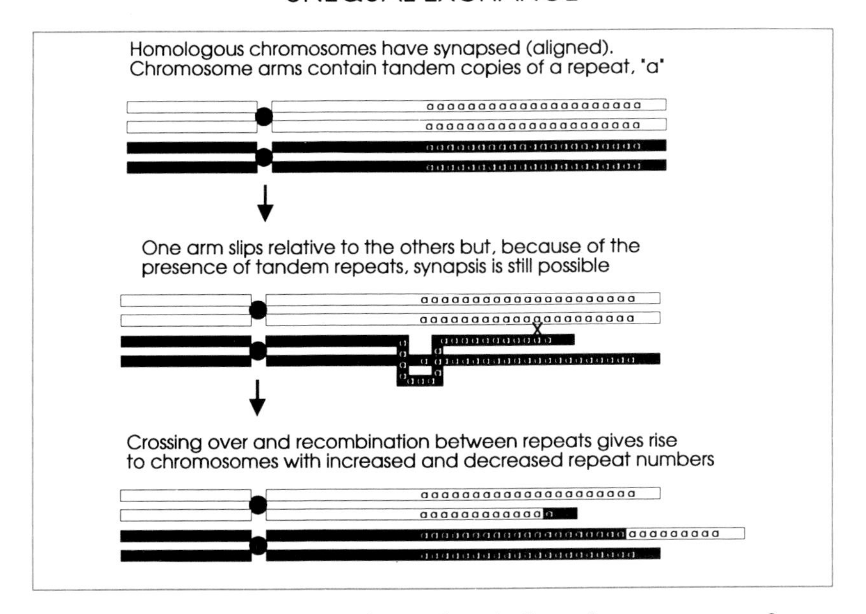

The most frequent source of mutations is from the movement of **transposable elements**. These are the so called "jumping genes". There are different classes of transposable elements or **TE**s. The most common group resembles retroviruses. They carry a gene that encodes the enzyme, **reverse transcriptase**, that catalyzes the conversion of a messenger RNA transcript of the element back into DNA. The new DNA copy can then insert into the host DNA at a new location from which new messenger RNA transcripts can be produced as a consequence of normal activity within the nucleus.

REPLICATION SLIPPAGE

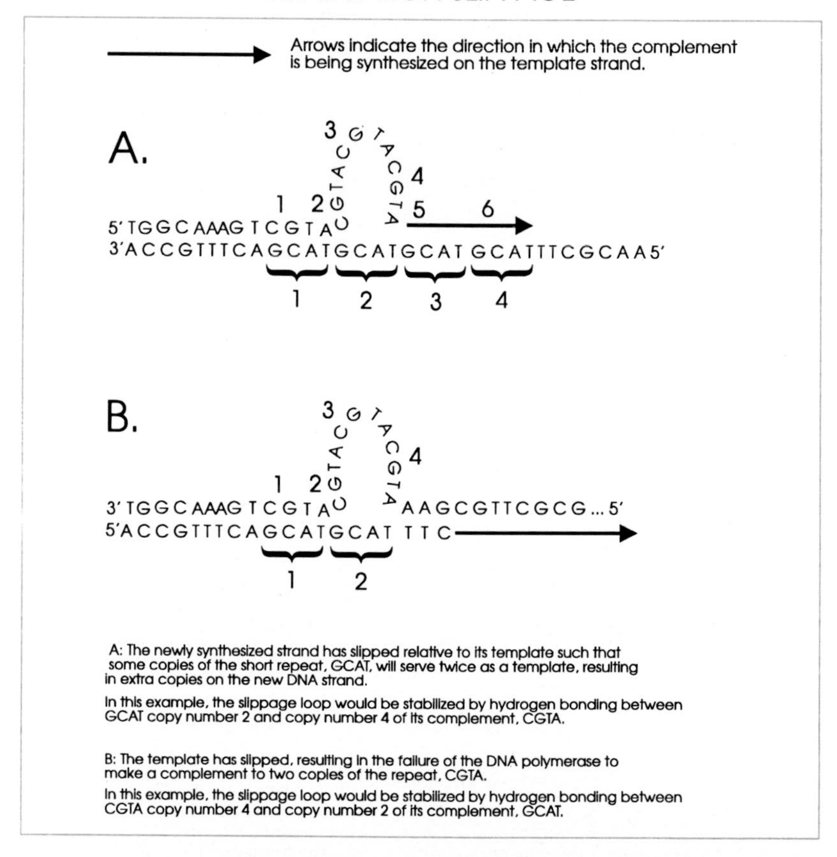

A: The newly synthesized strand has slipped relative to its template such that some copies of the short repeat, GCAT, will serve twice as a template, resulting in extra copies on the new DNA strand.

In this example, the slippage loop would be stabilized by hydrogen bonding between GCAT copy number 2 and copy number 4 of its complement, CGTA.

B: The template has slipped, resulting in the failure of the DNA polymerase to make a complement to two copies of the repeat, CGTA.

In this example, the slippage loop would be stabilized by hydrogen bonding between CGTA copy number 4 and copy number 2 of its complement, GCAT.

The insertion of transposable elements into new locations can alter the expression of neighboring genes, with profound phenotypic consequences. The movement of TEs may also lead to chromosome breaks and rearrangements. A **translocation** is the attachment of part of one chromosome to another while an **inversion** entails the reversal in the orientation of a segment of a chromosome. These mutations can have tremendous phenotypic consequences. In fact, among mammals, **morphological evolution**, that is, evolution of size and shape, is associated with such changes in karyotype. Because most genomes contain a large number of diverse TEs, it's possible they're a primary source of the mutations that provide the raw material for evolution.

A "JUMPING" GENE JUMPS

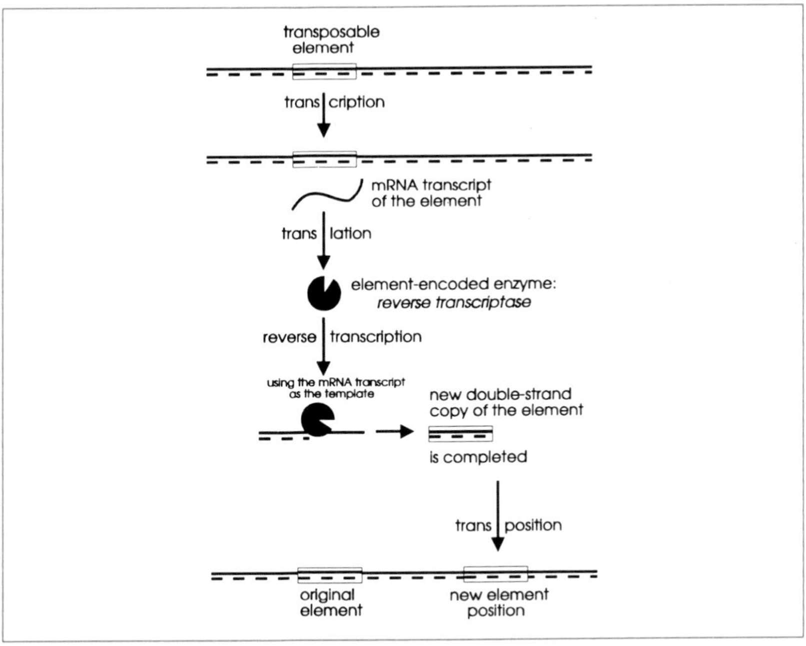

Mutations are not, however, the only source of genetic variation. Recombination, we have seen, can mix the nonalleles that are linked along a chromosome. Thus, recombination is an important source of variation by continually generating new combinations of nonalleles. Natural selection may favor certain combinations and select against others. Nevertheless, mutations are the ultimate source of genetic variation because they give rise to the different alleles present at each locus.

Now let's **review the key words**: point mutation, silent mutation, neutral mutation, frameshift, and transposable element. Point mutations affect a single nucleotide in the DNA code. Transitions convert one purine into another purine or convert one pyrimidine into another pyrimidine. Transversions convert purines into pyrimidines and *vice versa*. Substituting one base for another can cause the substitution of one amino acid for another in the encoded protein. If no amino acid change occurs, the mutation is silent. If a change occurs, but does not affect protein function, the mutation is neutral. Some point mutations cause the loss or gain of a single nucleotide.

A frameshift mutation occurs if this deletion or insertion takes place within a structural, or protein-specifying, gene. All codons of the messenger RNA transcript are misread if they occur after the point of the frameshift mutation in the corresponding DNA, drastically affecting the structure of the encoded protein. Many mutations, including deletions, insertions, and chromosomal breaks and rearrangements are the consequence of the movement of transposable elements, also called "jumping genes". Most of these resemble RNA viruses, called retroviruses, that encode an enzyme, reverse transcriptase, that converts RNA transcripts of the element into new DNA copies that can recombine into the DNA. Mutations provide the raw material for adaptive evolution.

CHAPTER 9

ON THE ORIGIN OF SPECIES

One of the most controversial aspects of Darwin's book, *On the Origin of Species*, is the emphasis placed on natural selection and consequent adaptation in the process of speciation. **Key words** for our discussion of speciation are species, gene flow, allopatric speciation, sympatric speciation, reinforcement, and punctuated equilibria. Natural selection leads to adaptation. Adaptations are traits that promote reproductive success in a particular environment for a particular organism. However, when one species gives rise to a new species, **speciation**, it's not clear that further adaptation is coincident. Some hypotheses about speciation emphasize adaptation while others reject it.

Darwin argued that speciation occurs by the slow, but steady, refinement and accumulation of adaptations. Darwin's thought was encouraged by **Charles Lyell**'s book, *The Principles of Geology*. Here, Lyell argued the case for **uniformitarianism**, the principle that the earth has always been subjected to the same slow, geological processes. Because of evidence that mountain ranges have risen and worn away, the earth must be very old. And, an old earth provided Darwin the time his model of speciation required. But, before discussing speciation, let's define species.

Most biologists accept the **biological species concept** of **Ernst Mayr**. Here, a **species** is defined as a set of populations that are actually, or potentially, capable of interbreeding and that are reproductively isolated from other such sets of populations. Populations are **reproductively isolated** if matings do not occur between their members or, if they do occur, do not lead to the production of fertile offspring.

Reproductive isolating barriers may be either premating or postmating in their operation. **Premating,** or **ethological**, isolating barriers prevent matings from occurring and include species specific signals used in **mate recognition**. These may be calls, songs, odors, dances, etc. **Postmating** barriers operate after a mating has occurred. They include the failure of sperm to fertilize eggs, the failure of embryos to complete development, the failure of the new born to survive to adulthood, and the infertility of the hybrid progeny.

A species is held together by the exchange of genes between its populations. Speciation is the evolution of reproductive isolation and entails the severing of this relationship between populations.

The most widely accepted model of speciation is divergence in allopatry or **allopatric speciation. Allopatry** means that two populations are isolated in space and cannot, therefore, exchange genes back and forth. Each population is then free to adapt to its own particular circumstances, perhaps with unique sets of mutations. As each adapts to its own environment, they grow farther and farther apart until at some point the two populations could not interbreed and exchange genes even if they came back into secondary contact. Allopatric speciation is more likely if the environments differ greatly and impart strong natural selection pressures and if the migrants that found new populations carry very little of the genetic variation that was present in the original population. Adaptation plays a large role in allopatric speciation. Darwin saw allopatric speciation as a consequence of the slow, steady accumulation of adaptations and their refinement in response to prevailing, local conditions.

Sympatric speciation, or the splitting of a single population into two, is a more controversial mode of speciation. **Sympatry** means that two species, populations, or subpopulations share the same environment. Sympatric speciation has two special requirements. First, individuals must vary in their ability to utilize two different resources or habitats in the common environment. Individuals good at utilizing resource 1 must be poor at utilizing resource 2, and *vice versa.* Second, individuals must mate at the preferred resource or prefer, as mates, those capable of using the same resource.

ALLOPATRIC SPECIATION

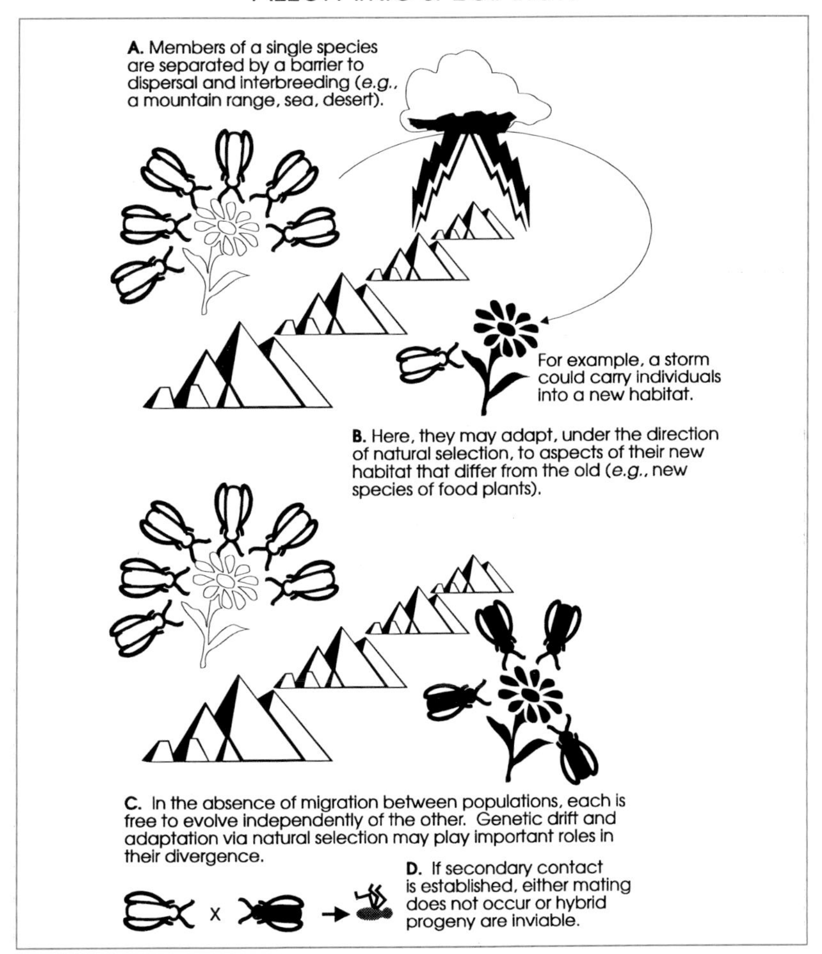

The first requirement leads to **disruptive selection**. Disruptive selection is a form of natural selection in which phenotypes, at either extreme, are favored while those in the middle are selected against. Here, individuals specializing on resource 1 and those specializing on resource 2 are favored and enjoy high reproductive success. However, generalists that specialized on neither resource, but that tried to exploit each, would have low reproductive success. For instance, imagine an insect population faced with two poisonous plants on which to feed. If adaptations to detoxify one poison interfere with adaptations to detoxify the other, survival requires feeding only on the plant with the poison that can be detoxified. Individuals feeding on both plant species would perish from the poison they could not detoxify. If phenotypes were arranged according to how much of each plant species was consumed, the extremes would be to eat only one species or the other. And, each of these extreme phenotypes would be favored by natural selection.

SYMPATRIC SPECIATION

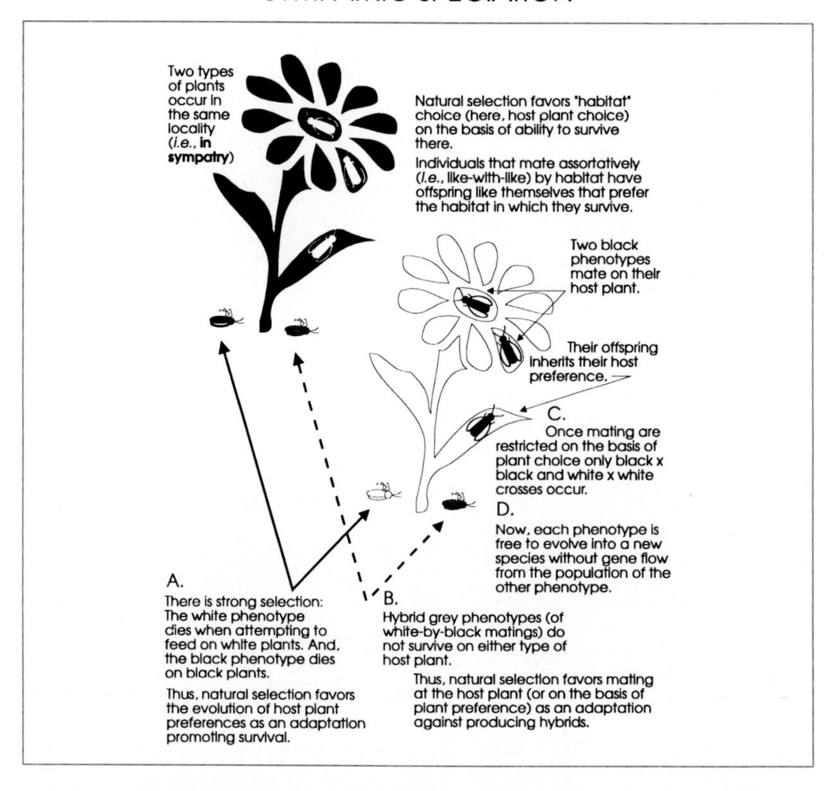

The other requirement for sympatric speciation is that mates share the same phenotype. This type of mating system is referred to as **assortative mating**. Assortative mating could result directly from mating preferences for one phenotype over another or indirectly from mating only at the preferred resource. In either case, assortative mating leads to the production of progeny that resemble their parents at the trait under disruptive selection and prevents the production of progeny of an intermediate phenotype.

MODES OF SELECTION

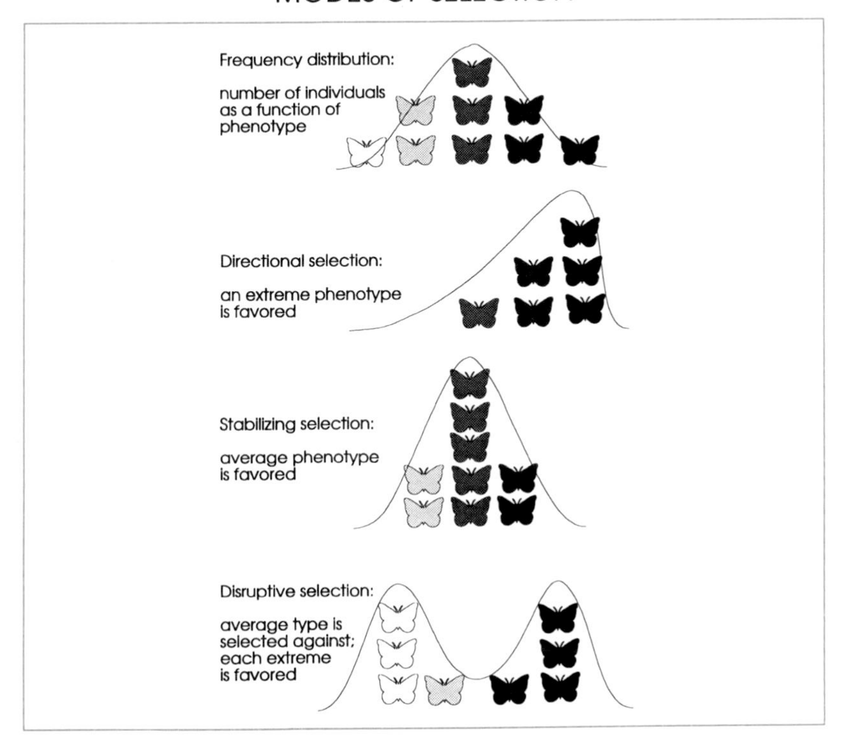

If disruptive selection is very strong, such that few, if any, individuals of intermediate phenotype survive to reproduce, and if mating is almost always assortative, sympatric speciation may occur. However, if disruptive selection or mating preferences are weak, each subpopulation will exchange genes with the other, preventing their ultimate divergence into different species. Because only the occassional transfer of genes can prevent sympatric speciation, it is generally considered to occur rarely, if at all. The transfer of genes between populations or subpopulations is called **gene flow**. Gene flow entails the movement of reproductive individuals, or their gametes, and can counter the impact of natural selection in promoting adaptation to local conditions.

Reinforcement for ethological isolation is a quasi-speciation mechanism first suggested by Alfred Russell Wallace. Here, two populations having evolved postmating barriers, while in allopatry, now establish secondary contact due to range expansion from one or both populations. Once in contact, individuals will interbreed because premating barriers are absent or weak. Yet, because postmating barriers are in place, mating efforts between them will be wasteful of time and energy. Therefore, natural selection favors the evolution of premating barriers because individuals that confine matings to members of their own population will have more time and energy to invest in viable, fertile offspring.

REINFORCEMENT FOR ETHOLOGICAL ISOLATION

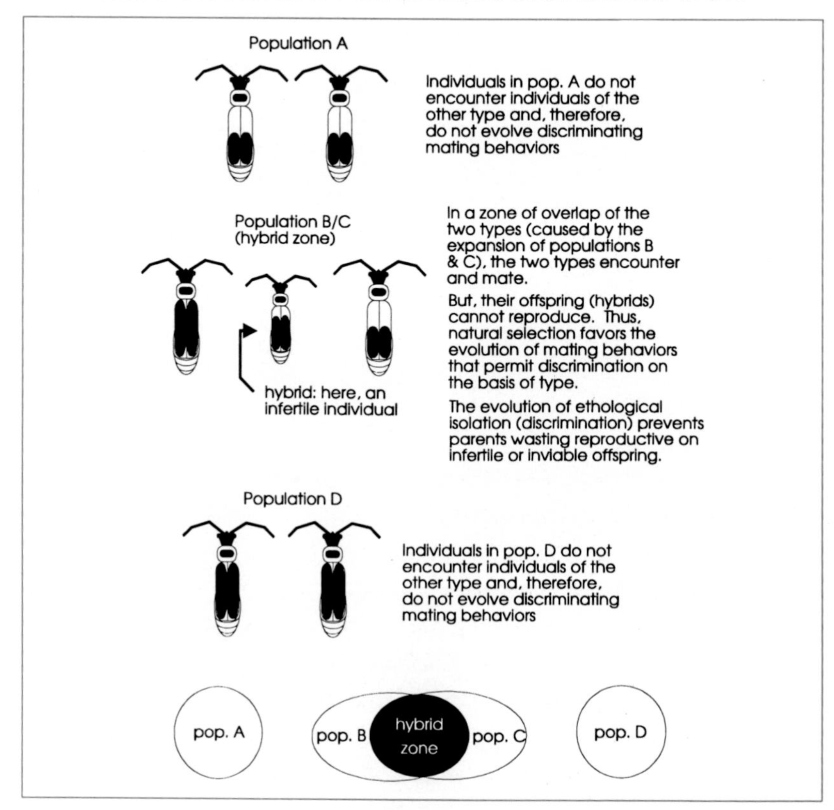

Reinforcement is the evolution of premating isolation as an adaptation to prevent gamete wastage when postmating barriers are already in place. If the postmating barriers are incomplete, that is, if some hybrid progeny survive and are fertile, reinforcement cannot evolve and, in fact, premating barriers erode with ensuing gene flow.

The theory of **punctuated equilibria** offers models of speciation that do not entail natural selection or ensuing adaptation. Here, natural selection produces adaptations to local conditions that refine the phenotype with respect to its environment. The result is microevolution, or change on a small scale only. Macroevolution, or gross morphological change, is envisioned to occur at the time of speciation. But, the large scale changes are not necessarily adaptive. Rather, phenotypic evolution is random in its direction. As a result, some species will be produced with morphological changes that favor species survival while others will be produced with morphological changes that promote extinction. What ensues is selection at the level of species, not Darwinian selection among individuals within a population. **Species selection** not natural selection is envisioned to determine trends in gross morphology.

For instance, consider the evolution of horses. The first horses were much smaller than modern horses and the size of horses has, for the most part, increased over evolutionary time. However, the fossil record does not reveal increases in size within a species over evolutionary time. Rather, as new species arose, some were larger and some were smaller than those already present. Over time, the species that survived longest or that gave rise to the greatest number of daughter species were, primarily, those with larger sized individuals. Consequently, species selection may account for the evident trend that size increased. This contrasts with the classical view of directional natural selection acting within populations.

Directional selection occurs when only one extreme is favored among the range of phenotypes present. Directional selection was long held to account for the evolution of larger size among horses. Suppose, for example, larger horses could more readily escape large predators such as sabre-toothed cats. Thus, within a population, larger individuals would have left more offspring than smaller individuals because they were less likely to fall prey to predators. From generation to generation and through long periods of time, size would increase, via natural selection, as mutations contributed more genetic variation for size.

The fossil record is purported to support the theory of punctuated equilibria and attendant species selection. For instance, morphology is constant within fossil species for long periods of time. This condition is called **stasis** and represents an equilibrium condition. Morphological change appears to burst forth at the time of speciation. The equilibrium is **punctuated** by a period of brief, but rapid, change. Adherents of the punctuated equilibria theory postulate that speciation occurs in very small isolated populations that become **fixed** through genetic drift (see pp. 77-78) for chromosomal mutations, such as inversions or translocations, that both alter morphology and cause reproductive isolation from the ancestral populations. An allele or chromosomal aberration is said to be fixed when all individuals in the population possess that allele or aberration. Alternative alleles or chromosomes are then extinct.

PUNCTUATED EQUILIBRIUM/SPECIES SELECTION

Horse species "a" arises in time period 1. In time period 2, "a" gives rise to two new species, "b" and "c".

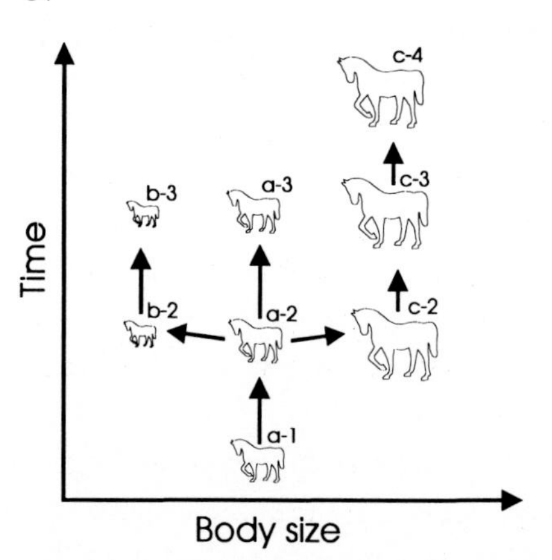

Species "a" does not evolve throughout its history (time periods 1-3) but new species with very different morphologies (body size, here) evolve rapidly in time from "a". This is the essence of a "punctuated equilibrium".

Species "a" gives rise to both smaller ("b") and larger ("c") sized species but only the larger sized "c" survives to period 4. Without a complete fossil record, we could incorrectly infer an evolutionary trend of increased size (for example, comparing a-1 with c-4)

Let's **review the key words**: species, gene flow, allopatric speciation, sympatric speciation, reinforcement, and punctuated equilibria. A species is a set of populations that are actually, or potentially, united by the back and forth exchange of genes. Gene flow refers to the movement of genes between populations as a result of the migration of individuals or the dispersal of their gametes. A species is reproductively isolated, or incapable of exchanging genes, with other species. Speciation is the evolution of reproductive isolation in one or more populations within a species. Allopatric speciation occurs between populations isolated by barriers to migration and gamete dispersal. Reproductive isolation evolves as a byproduct of adaptation to local conditions. Sympatric speciation is the splitting of a single population into two reproductively isolated units. It requires strong disruptive selection along some environmental parameter and also requires strong assortative mating with regard to the trait experiencing disruptive selection. Reinforcement is the sympatric evolution of premating barriers to the exchange of genes, such as species-specific mating signals, after postmating barriers, such as sterility of hybrid offspring, have evolved in allopatry. The theory of punctuated equilibria accounts for speciation without adaptation and views species as relatively stable in morphology. Speciation then occurs in small, isolated populations where drift can effectively oppose natural selection, leading to rapid morphological evolution during a brief speciation period.

CHAPTER 10

POPULATION FITNESS

Natural selection leads to adaptation, an increase in the frequency of traits that increase relative reproductive success. Note that natural selection occurs between members of an interbreeding population. Thus, any trait that increases reproductive success relative to other traits possessed by population members will be favored. It is irrelevant that reproductive success may still fall far short of that of members of other populations or species.

Humans and house flies may be equally adapted to their own respective environments. Clearly, humans do not represent the pinnacle of evolutionary adaptation. The human traits we value most, intelligence, for example, may not enhance the reproductive success of a house fly. Our intelligence is an adaptation to the environment of our ancestors and does not make us better than less intelligent organisms, only different.

The selection for adaptations is driven only by the variation in reproductive success within a population. **Key words** for our discussion of this variation are: genetic drift, modes of selection, fitness, selection coefficient, sexual selection, and frequency dependent selection. The reproductive success of genotypes within a population is **absolute fitness**, symbolized by **W**. **Relative fitness**, symbolized by **w**, is the absolute fitness of a genotype divided by the absolute fitness of the most fit genotype. Theoretically, relative fitness can vary from 0 to 1. The higher the relative fitness of a genotype, the more it contributes alleles to the next generation. The strength of natural selection is indexed by the **selection coefficient**, symbolized by a **s**, which is simply calculated as 1 - relative fitness. The higher the selection coefficient, the less a genotype contributes alleles to the next generation. Selection coefficients also vary from 0 to 1. **Mean population fitness** is calculated as the genotype frequency multiplied by absolute fitness, summed across all genotypes within the population. Although mean, or average, population fitness always increases as a consequence of natural selection, other evolutionary forces, such as genetic drift, may have a negative impact on mean fitness.

GENETIC DRIFT

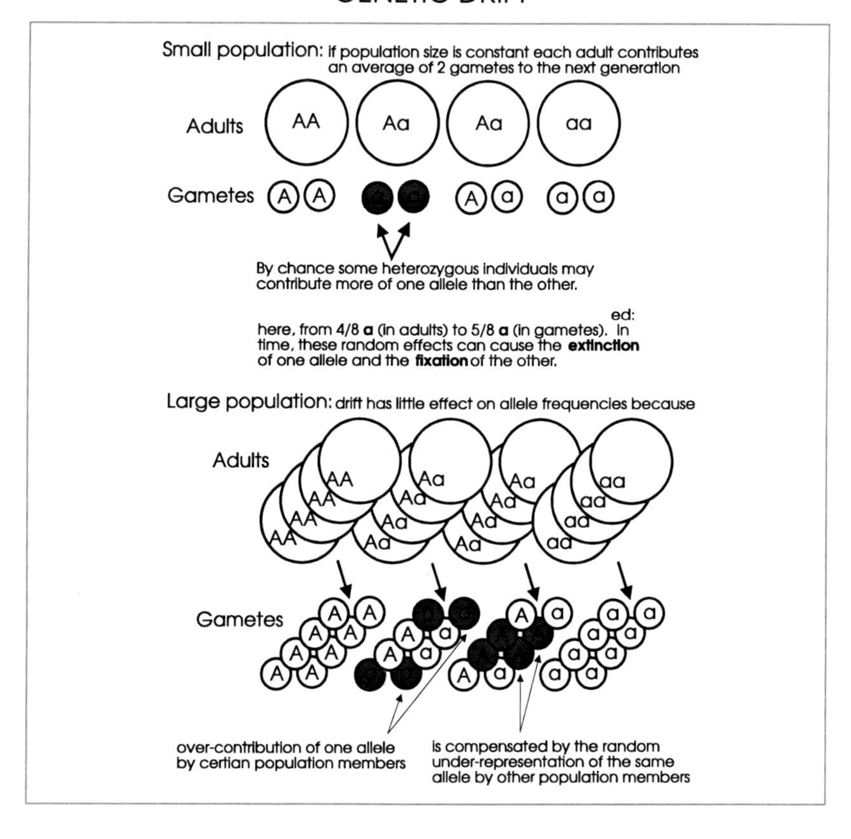

Genetic drift is a change in allele frequencies that can occur when the next generation is established from a small number of gametes. With smaller samples, it becomes increasingly likely that allele frequencies in gametes will not match those in the population at large. Genetic drift in small populations can overwhelm the influence of natural selection. A small amount of gene flow can also retard, or reverse, the impact of natural selection within a population. **Gene flow** is the movement of gametes or reproductive individuals from one population into another. Gene flow can prevent adaptation to local conditions by reintroducing alleles that contribute to less fit genotypes. However, gene flow could, just as well, introduce favorable mutations, providing the genetic variation on which the evolution of adaptation depends.

Genetic drift can also have a positive effect on the mean fitness in a population. For instance, suppose that the genotype **AABB** is favored over all genotypes except **aabb**. Now suppose the frequencies of the **a** and **b** alleles are each very low because they are new mutations. The odds are against either allele rising in frequency due to natural selection because their rarity ensures that they will rarely occur together in the same genotype. Rather, one or the other occurs in less fit genotypes, such as **AABb** and **AaBB**. However, genetic drift could overwhelm natural selection in small populations, causing the frequencies of **a** and **b** to rise. Now, the most favored genotype is likely to occur providing for favorable selection on the new alleles that results in a higher mean population fitness.

One form of selection can actually decrease mean population fitness. This is **sexual selection**; the change in trait frequency due to differences in mating success. Darwin introduced the concept in 1871 in his book, *The Descent of Man and Selection in Relation to Sex*. In most species, sexual selection operates more strongly in males because males do not invest as much as do females in producing and rearing offspring. Consequently, reproductive success in males often depends primarily on the number of females mated while in females reproductive success depends primarily on the number of energetically expensive offspring produced or reared.

Darwin argued for two types of sexual selection, intersexual and intrasexual. **Intersexual selection** arises from differences in the attractiveness of males to females while **intrasexual selection** arises from differences in the ability of males to compete among themselves for access to females. Intersexual selection may represent the female's choice for "**good genes**" that, for instance, confer resistance to parasites. Female choice may have selected for traits such as the bright, colorful feathers in birds that permit males to broadcast their health. Intrasexual selection in males may have promoted the evolution of aggressive and territorial behavior, and large canines, tusks, and antlers.

GENETIC DRIFT & POPULATION FITNESS

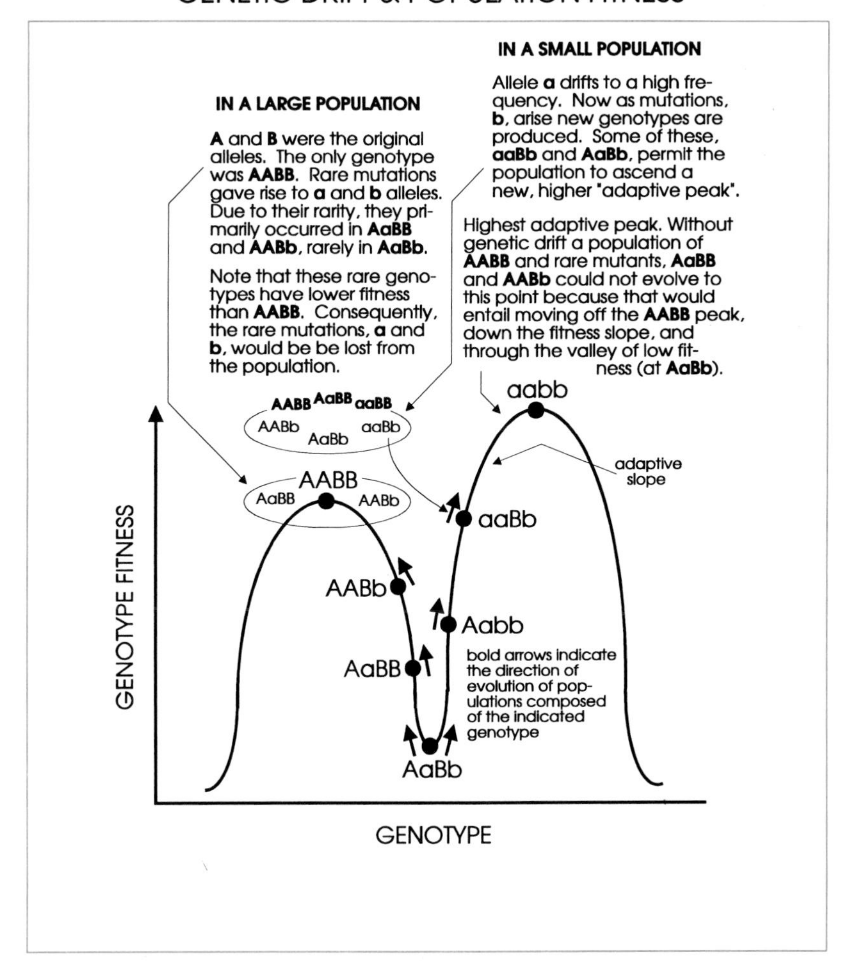

Sexual selection can favor traits that promote mating success but compromise adaptation under natural selection. For instance, sexual selection might favor larger size for enhanced fighting ability. But, the attainment of large size would require prolonging the juvenile period, resulting in an increased risk of predation or starvation. As these risks increase, mean population fitness decreases. A balance may eventually be struck that represents the conflicting pulls of natural and sexual selection on the phenotype.

Whereas sexual selection is often directional, favoring more and more extreme manifestations of a trait such as body size or tusk length, natural selection is frequently stabilizing. **Stabilizing selection** favors some intermediate degree of expression and disfavors both extremes of the phenotypic distribution. For instance, consider gestation length. If it's too long, birth is difficult and the lives of the mother and her offspring may be threatened. On the other hand, if the gestation period is too short, the offspring may be too small and underdeveloped.

FREQUENCY DEPENDENT SELECTION

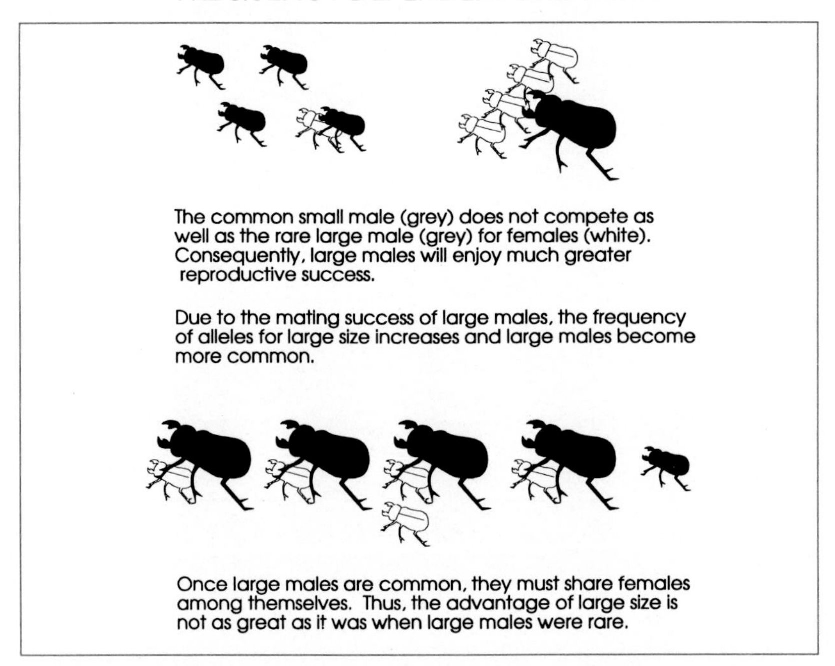

Sexual selection is also frequency dependent. **Frequency dependent selection** means that the absolute fitness of a genotype decreases as its frequency increases. For instance, imagine two types of males, one attractive and one unattractive to potential mates. In a population composed mainly of unattractive males, the rare attractive type would enjoy great reproductive success because there would be many females per attractive male. However, generations later, the frequency of attractive males would be much higher. Therefore, the number of females per attractive male would be lower. It's always better to be attractive than not, but the absolute fitness of the attractive type, continues to decline as its numbers increase.

Now let's **review the key words**: genetic drift, modes of selection, fitness, sexual selection, and frequency dependent selection. Genetic drift is the random change in allele frequencies that occurs in small populations because, only a small number of gametes are used to establish the next generation. Consequently, it's unlikely that allele frequencies in gametes will exactly match those of the individuals producing the gametes. There are three primary modes of selection: directional, disruptive, and stabilizing. Directional selection favors one or the other extreme of phenotypes present in the population. Thus, the average value for the trait changes over time, becoming larger or smaller. Disruptive selection favors both extremes with intermediate phenotypes having the lowest reproductive success. Stabilizing selection favors intermediate phenotypes. Consequently, for a trait such as body weight, the average value does not change but the variance or range of weights decreases as very heavy and very light phenotypes disappear. Fitness is the reproductive success associated with a genotype. Relative fitness is calculated as the reproductive success of a genotype divided by the reproductive success of the best genotype. Relative fitness varies from 0 to 1. The selection coefficient is calculated as 1 - relative fitness and is a measure of the force of selection against a genotype. Natural selection causes the average fitness of a population to increase. However, one form of selection, sexual selection can oppose natural selection and may cause average fitness to decline. Sexual selection arises from variation in mating success and favors the traits in males that increase their attractiveness to females or increase their ability to fight other males for access to females. Sexual selection is frequency dependent meaning the fitness of a genotype is lower the more common the genotype.

CHAPTER 11

UNICELLULAR & SIMPLE MULTICELLUAR ORGANISMS

All life on earth is believed to have evolved from a common unicellular organism that, itself, evolved about three billion years ago. Consequently, similarities among bacteria, plants, fungi, and animals in cellular structure and biochemistry are reflections of common ancestry. Likewise, all vertebrates share a common invertebrate ancestor accounting for similarities in body plan. Thus, the classification of organisms reflects current opinion about evolutionary history. Evolutionary theory unifies the science of biology.

Over one million organisms are currently classified into five broad or very inclusive groupings, called **kingdoms**. **Key words** for our discussion of these kingdoms are: Monera, Protista, Fungi, Plantae, and Animalia. All the organisms of these kingdoms share certain features not present in other kingdoms. The kingdom **Monera** consists of **bacteria** and **cyanobacteria**. Cyanobacteria are also called blue-green algae. Bacteria and cyanobacteria lack membrane-bound intracellular organelles such as mitochondria, chloroplasts, the endoplasmic reticulum, and the nucleus. They're also called

prokaryotes because they lack nuclei. Their ribosomes are very similar to those found within mitochondria and chloroplasts but differ in RNA and protein composition from those in the cytoplasm of eukaryotes, organisms with a nucleus. Bacteria may possess whip-like **flagella** for locomotion. Here, a flagellum is composed of the protein **flagellin** and is attached to the cell surface.

Bacteria are a very diverse group. **Archaebacteria** are characterized by the absence of **peptidoglycan** in their cell walls, whereas **Eubacteria** possess peptidoglycan. The archaebacteria includes the **methanogens** that produce swamp gas by using hydrogen gas to reduce carbon dioxide to methane, CH_4. Methanogens are anaerobic. They live in the absence of oxygen. Their habitats include marsh and swamp muds, and the intestinal tracts of animals. Other archaebacteria are the **extreme halophiles** that require a very high salt environment and the **thermoacidophiles** that live in habitats with temperatures of 60-80°C and an acidic pH of 2-4. Archaebacteria may be more closely related to eukaryotes than are eubacteria, as assessed by their ribosome structure.

RELATIONS AMONG BACTERIA

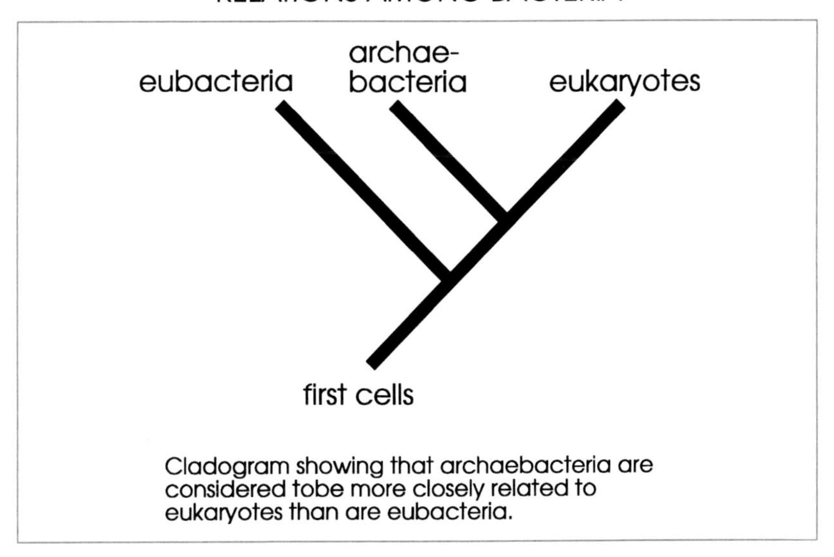

Cladogram showing that archaebacteria are considered tobe more closely related to eukaryotes than are eubacteria.

The eubacteria are even more diverse than archaebacteria but inhabit less extreme environments. **Actinomycetes** are colonial branching bacteria of the soil. **Chemoautotrophic bacteria** derive energy from the oxidation of inorganic, as opposed to organic, compounds. Photosynthetic **cyanobacteria** may form filamentous colonies that contain specialized **nitrogen-fixing** cells called **heterocysts. Endospore forming bacteria** survive harsh conditions as a spore; a dehydrated cell within a protective encasement. **Enteric bacteria** are anaerobes that live within animal intestinal tracts. **Mycoplasmas** lack cell walls and live as saprobes, feeding on decaying plant and animal matter, or live as pathogens. **Myxobacteria** are soil dwellers that move using gliding motions. Under harsh conditions, they disperse as spores from a stalk produced by the aggregation of individual cells. Some **nitrogen- fixing bacteria** form nodules on the roots of legumes where inorganic nitrogen is captured in the production of essential organic metabolites. Common bacteria in marine and aquatic sediments are the **phototrophic anaerobes** that use H_2S, instead of H_2O, as an electron source in photosynthesis. **Pseudomonads** are common in many aquatic and soil habitats. **Rickettsias** and **chlamydias** are intracellular, or within cell, parasites of animals. And, **spirochaetes** are very long helical bacteria that live as saprobes or parasites.

The kingdom **Protista** consists of single celled, eukaryotic organisms. Most protists possess either **flagella** or shorter, more numerous, **cilia** for locomotion. Cilia also aid in the capture of food. The flagella and cilia of protists and other eukaryotes are composed of a circular array of nine double protein tubules that are contained within extensions of the cytoplasm. Photosynthetic, plant-like protists are called **algae** while ingestive, animal-like protists are called **protozoa**.

There are seven phyla of algae. The **phylum**, or phyla for plural, is the category or level of classification just beneath that of kingdom. Algal phyla are distinguished by their photosynthetic pigments, carbohydrate food reserve, number and location of flagella, and their cell wall components. For illustrative purposes we'll focus on the cell wall. **Dinoflagellates**, phylum **Dinoflagellata**, are marine and freshwater algae without an external cell wall. They do, however, possess submembrane **cellulose**, the primary component of plant cell walls. **Golden algae**, phylum **Chrysophyta**, are primarily fresh water algae with cell walls composed of the polysaccharide, **pectin**, reinforced with siliceous compounds. **Diatoms**, phylum **Bacillariophyta**, are inhabitants of marine and fresh waters and possess walls of hydrated **silica** within an organic matrix. The relatives of the common fresh water organism *Euglena*, phylum **Euglenophyta**, have no cell wall but do possess a submembranous layer of protein. **Green algae**, phylum **Chlorophyta**, are primarily fresh water organisms with **cellulose** cell walls. **Brown algae**, phylum **Phaeophyta**, primarily inhabit cold marine waters. Brown algae possess a cell wall of cellulose and other polysaccharides. Finally, the **red algae**, phylum **Rhodophyta**, consists mainly of tropical marine species with a cell wall of cellulose mixed with other polysaccharides.

Animal-like **protozoans** possess a specialized organelle, the **water vacuole**, that expells excess osmotically-derived water. The **Rhizopoda** are **amoebas**, either naked or shelled, with broad flowing extensions of the cytoplasm called **pseudopodia** that function in movement and capture of food. The **Actinopoda** are species with slender, radiating pseudopods. The Actinopoda includes **radiolarians** that have an **internal skeleton** and **heliozoans** that do not. Many protozoans that are **parasitic** in animals are classified in the phylum **Apicomplexa**. The **Zoomastigophora** use **flagella** for locomotion and prey capture. Some species are colonial. **Ciliated protozoans**, phylum **Ciliophora**, including members of the genus *Paramecium*, use cilia for locomotion and the capture of prey.

CILIATED PROTOZOAN

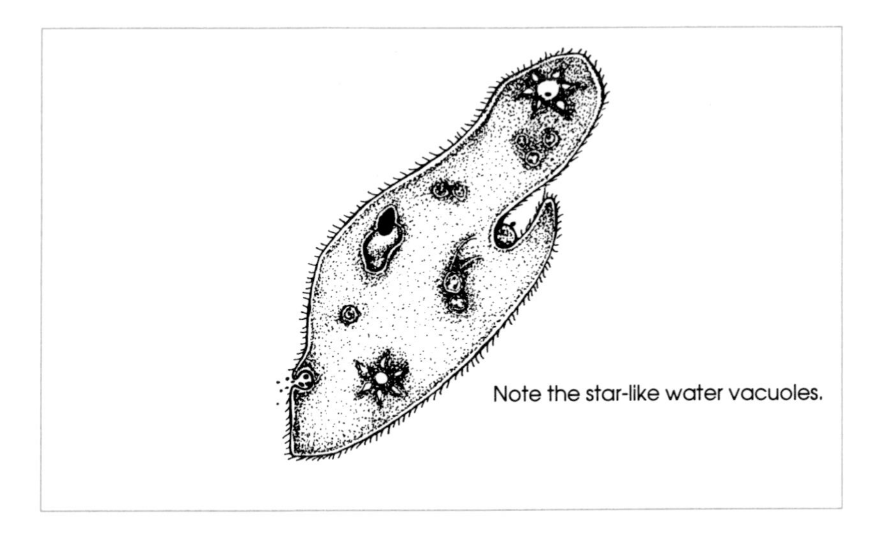

Note the star-like water vacuoles.

Ancient protistans are believed to have given rise to the other eukaryotic kingdoms. For instance, members of the phylum **Chytridiomycota**, which includes free-living saprobes and parasites, are very similar to fungi. Both possess a hyphal, or branched, structure and cell walls of **chitin**. Other **fungus-like** protists include plasmodial, or acellular, **slime molds**, phylum **Myxomycota**, cellular slime molds, phylum **Acrasiomycota**, and **water molds**, phylum **Oomycota**.

CHAPTER 12

FUNGI AND PLANTS

The kingdom **Fungi** includes primarily **multicellular** organisms which possess **cell walls** composed of **chitin**. The body plan of fungi is characterized by a series of filamentous, tubular **hyphae** that are organized into a network called a **mycelium**. Cells of the hyphae are **haploid** but may be genotypically diverse due to the fusion of hyphae of different individuals. In **coenocytic fungi**, the hyphae are **aseptate**, meaning the cells are not separated from each other by cell walls. Some of the hyphae of parasitic fungi are modified into tissue penetrating **haustoria**.

There are three major divisions of fungi. Each division is characterized by features of **plasmogamy**, the fusion of the cytoplasm of sex cells, by features of **karyogamy**, the fusion of nuclei of sex cells, and by the length of time a mycelium exists as a dikaryon. The cells of a **dikaryon** contain two, paired nuclei that have not fused. Dikaryotic mycelia may persist for years before karyogamy occurs. The unfused nuclei represent the original contributions of the sex cells.

The division **Zygomycota** includes primarily soil saprobes and species that form mycorrhizae with plant roots. **Mycorrhizae** are mutualistic associations, between a fungus and a plant, that permit the exchange of nutrients and metabolites. Sexual **zygospores** remain dormant for long periods in draught-resistant **zygosporangia**.

LIFE CYCLE OF A FUNGUS

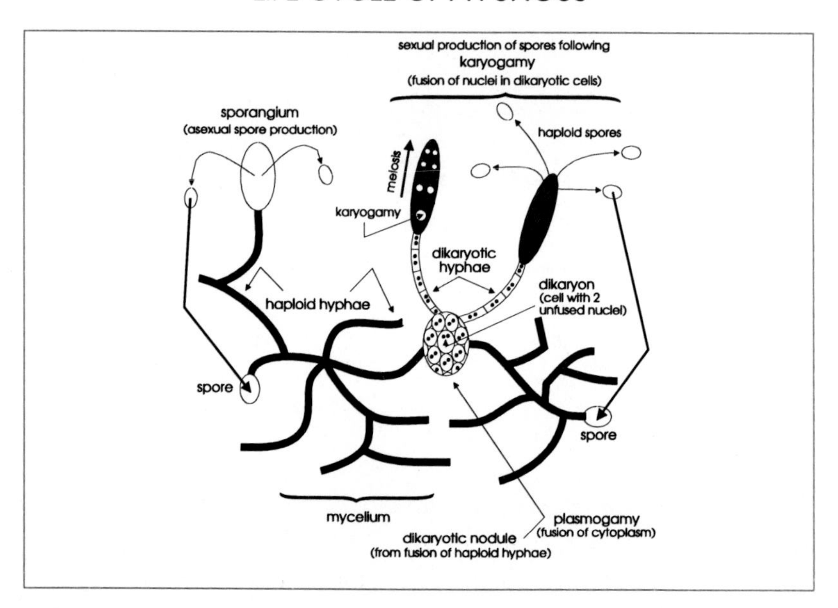

Yeast, **cup fungi**, and **truffles** are members of the division **Ascomycota**. While some **ascomycetes** are plant pathogens or saprobes, **lichens** are the product of a mutualistic relationship with algae. Ascomycetes produce asexual spores from hyphal extensions called **conidia**. These give rise to haploid mycelia. Within a mycelium sexual structures arise. When two such structures fuse, they produce a dikaryotic nodule, the **ascogonium**. The ascogonium gives rise to hyphae that branch out and bear sac-like **asci**. Hence the common name, sac fungi. Within each **ascus**, karyogamy gives rise to diploid cells that undergo meiosis to produce **haploid ascospores** that are released and may germinate to give rise to new mycelia.

The other division of fungi is **Basidiomycota** which includes **mushrooms, puff balls** and **rusts**. Sexual reproduction entails the fusion of two hyphae that then give rise to a dikaryotic mycelium. This grows into a fruiting structure, the **basidiocarp**, of which the mushroom is an example. **Gills** of the basidiocarp are lined with club-like structures called **basidia**.

Karyogamy and meiosis occur within each **basidium**, giving rise to haploid **basidiospores** which can establish new haploid mycelia.

Previously, fungi were classified with plants and were considered to represent either primitive forms or degenerate plants that had lost the capacity to produce **photosynthetic chlorophyll**. Currently, only eukaryotic, multicellular, photosynthetic autotrophs are classified as plants. An **autotroph** is an organism that derives energy from sunlight or the oxidation of inorganic chemicals and not, as do **heterotrophs**, from the consumption of other organisms, alive or dead. Traditionally, as with fungi, plants have been organized into divisions not phyla. For instance, **nonvascular plants**, those without tissues for the transport of water and nutrients, are classified into three divisions. These are **Bryophyta**, or mosses, **Hepatophyta**, or liverworts, and **Anthocerophyta**, or hornworts.

Vascular plants have xylem and phloem for the transport of water and nutrients. In flowering plants, **xylem** conducts water and minerals from the soil. It's composed of elongate cells called **tracheids** that are dead at maturity. Water passes through the thin cell walls that separate these individual cells. Xylem also consists of living cells called **vessel members**. Water flows through openings in the cell walls of these living cells. **Phloem** transports the sugars produced by photosynthesis. Both its cell types, sieve tube members and companion cells, are alive at maturity. **Sieve tube members** pass sugars via cytoplasmic connections that extend through holes in their cell walls. **Companion cells** transfer sugars to and from the sieve tube members.

Vascular plants may be **seedless** or seed-producing plants. Ferns, division **Pterophyta**, are familiar seedless plants. Other seedless plants are: club mosses, division **Lycophyta**; horsetails, division **Sphenophyta**; and whiskferns, division **Psilophyta**. **Seed-producing** plants are classified into two groups: gymnosperms and angiosperms. Gymnosperms evolved long before angiosperms. Ancient **gymnosperms** include cycads, division **Cycadophyta**, and ginkgos, division **Ginkgophyta**. The seeds of gymnosperms do not develop within enclosed chambers as is the case with angiosperms. The most familiar gymnosperms are pines, firs, spruce, junipers, redwoods, and cypresses, all of which are members of the division **Coniferophyta**. Conifers are **cone-producing** evergreens.

Angiosperms are flowering plants. **Flowers** develop at shoot tips where they're enclosed within the **calyx**, a series of leaf-like **sepals**. Interior to the sepals are **petals**. Collectively, sepals and petals constitute the **corolla**. Inside the corolla are **stamens**, slender stalks capped with two-lobed **anthers** that each contain four **pollen sacs**. The inner most part of a flower is the **carpel**, or female part. There may be multiple carpels per flower. The lower part of a carpel is the **ovary** where eggs develop and seeds mature. The upper part consists of a stalk, the **style**, which bears a swollen **stigma** for the receipt of pollen. **Perfect flowers** have both carpels and stamens; **imperfect flowers** have one or the other.

FLOWER STRUCTURE

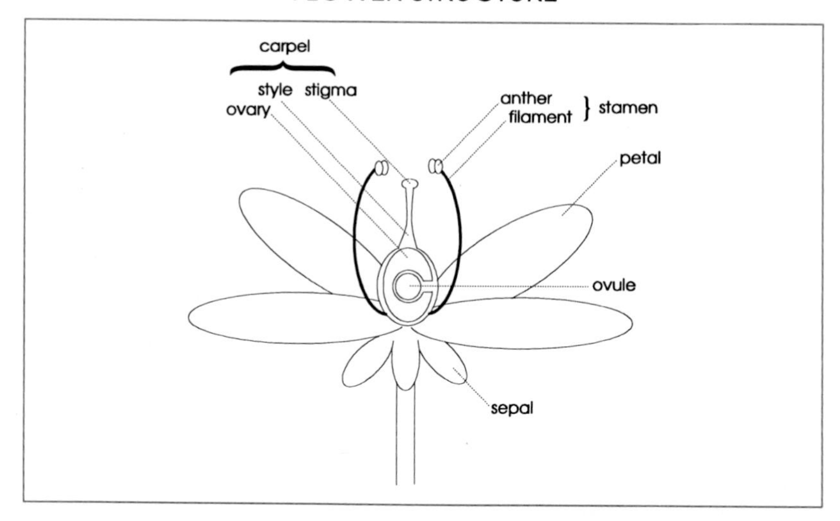

The body of a flowering plant is called the **sporophyte**. Eggs arise from cell clusters called female **gametophytes**. Within the ovary, the ovule begins as a cell mass attached to the ovary wall by a stalk. The inner tissue of the cell mass is the **nucellus** which is surrounded by either one or two integuments. A diploid **mother cell** within the nucellus undergoes meiosis to produce four haploid **megaspores** of which only one survives. The survivor undergoes three rounds of mitosis. Following cytoplasmic division there are seven cells, one becomes the haploid egg, and one diploid cell forms the nutritive **endosperm**. Both the egg and the diploid cell are later fertilized by haploid pollen derived nuclei. The now diploid embryo absorbs nutrients from the triploid endosperm via **cotyledons**, or seed leaves. The integuments later form a **seed coat**. A **fruit** is the matured ovary and associated flower parts.

Male gametophytes are the **pollen grains**. They form in four pollen sacs within the anther where a **diploid microspore** produces a mass of mother cells by mitosis. Each mother cell undergoes meiosis to produce **haploid microspores**. These then undergo mitosis to produce two-celled haploid pollen grains.

Two broad groupings of flowering plants are monocots and dicots. **Monocots** include grasses, lillies, irises, orchids, palms, and corn. They're characterized by flowers with sepals and petals in multiples of three, parallel leaf veins, blade-like leaves that arise from a **stem sheath**, and dispersed vascular tissues. **Dicots** include hard wood trees and asters. They have flower parts in multiples of four or five, net-like leaf veins, leaves that attach to the stem via a **petiole**, and vascular tissue organized into a ring of bundles.

CHAPTER 13

ANIMALS

In plants and fungi, the division is the taxonomic unit that corresponds to the grouping, the **phylum**, in animals. **Phylogeny** is the relationship between taxonomic groups that reflects evolutionary history. At the level of phyla, extant (that is, living) animals can be grouped according to characteristics either of the body plan or of embryonic development. For instance, the **Parazoa** lack true tissues while the **Eumetazoa** possess true

PARAZOANS, SPONGES

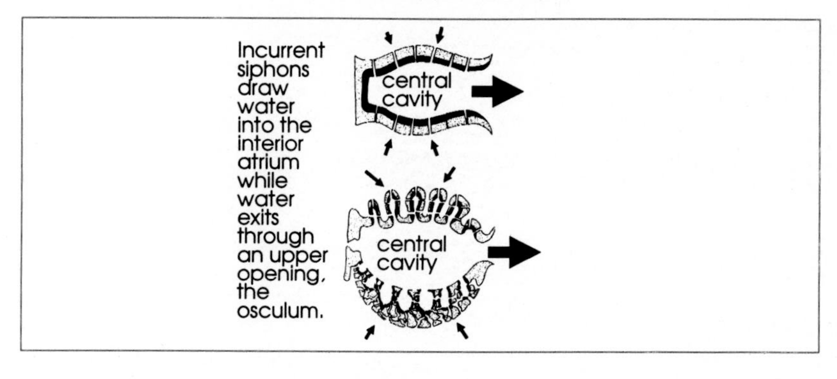

body tissues. The only extant parazoans are sponges, members of the phylum **Porifera**. Eumetazoans are broadly defined by body their **symmetry**. Comb jellies, phylum **Ctenophora**, and jellyfishes, corals, and anemones, all of the phylum **Cnidaria**, are **radially symmetrical**, meaning the body is composed of similar parts that are identically arranged around a central axis. This grouping of phyla is the **Radiata**. Other animal phyla, the **Bilateria**, are **bilaterally symmetrical**, meaning there are left and right sides that are the mirror images of each other. Bilateral symmetry is associated with **cephalization** or the concentration of sensory organs in a head region.

A CTENOPHORE, A MEMBER OF THE RADIATA

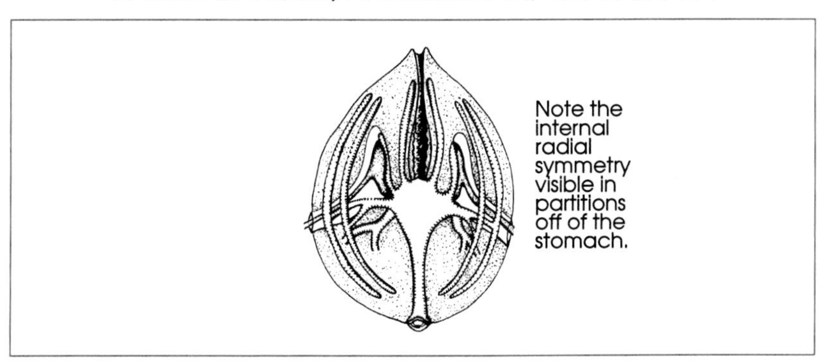

RADIALLY SYMMETRICAL CNIDARIANS

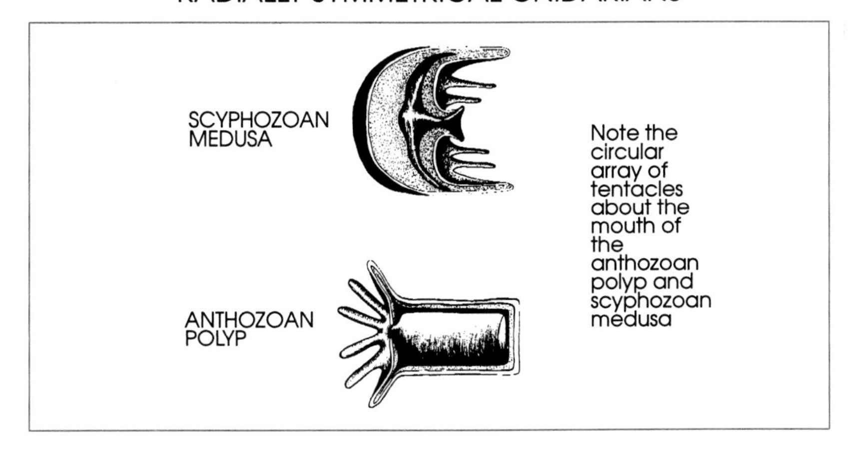

Associations among phyla of the Bilateria are based on events in **early embryonic development**. The embryo first develops into a ball of cells, the **blastula**, within which is a cavity called the **blastocoel**. As cells continue to divide mitotically, a depression on the surface of the blastula develops and sinks into the blastocoel as a pocket. The embryo is now called a **gastrula** and resembles a balloon with part of the outer surface pushed into the interior space. The outer surface cells are referred to as the **ectoderm** while those lining the pocket are called the **endoderm**. The space within the pocket is the primitive gut or the **archenteron** and its opening at the surface of the gastrula is the **blastopore**. A third cellular layer, the **mesoderm**, forms in the blastocoel between the ectoderm and endoderm. The mesoderm may arise as outpockets of the archenteron.

BODY PLANS

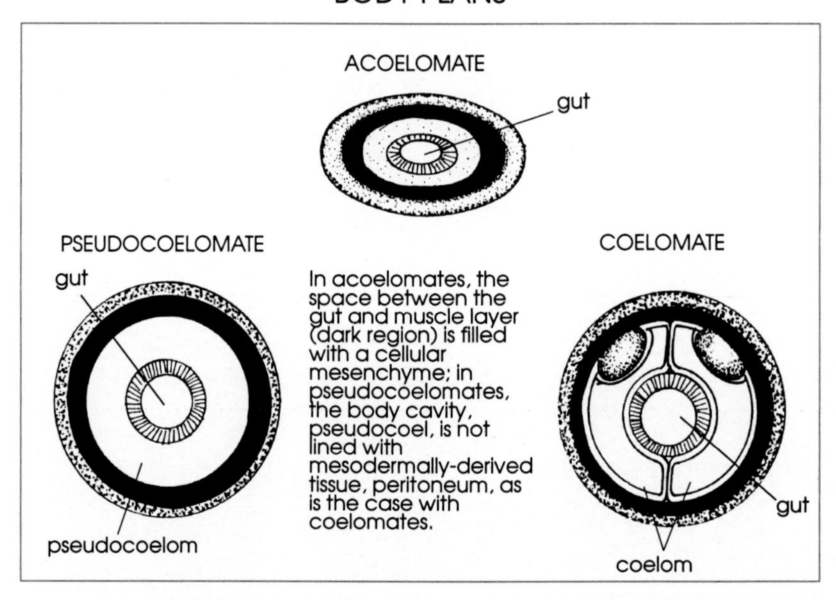

In acoelomates, the space between the gut and muscle layer (dark region) is filled with a cellular mesenchyme; in pseudocoelomates, the body cavity, pseudocoel, is not lined with mesodermally-derived tissue, peritoneum, as is the case with coelomates.

The endoderm gives rise to the lining of the gut and organs derived from the gut such as the liver and lungs. Ectoderm gives rise to the outer body covering and the central nervous system. Mesoderm gives rise to muscles and any other internal organs. **Acoelomates** have no cavity between the intestinal tract and the body wall. Extant acoelomates are flatworms, members of the phylum, **Platyhelminthes. Pseudocoelomates** have a body cavity but it's not lined with tissue of mesodermal origin. Pseudocoelomate phyla are **Rotifera**, the rotifers, and **Nematoda**, the roundworms. Pseudocoelomates and acoelomates may be relatively closely related. Additionally, both of these groups may be closely related to protostomes, a major grouping of coelomate phyla.

A PSEUDOCOELOMATE - A PLANARIAN (FLATWORM)

Note the upward facing eyes and the branching digestive system.

Coelomates are dichotomized into protostomes, or **schizocoelomates**, and deuterostomes, or **enterocoelomates**. All coelomates have a body cavity lined with tissue of mesodermal origin but the coelom of **protostomes** develops as a cavity within a solid mass of mesoderm while in **deuterostomes** the coelom arises from mesodermal outpocketing of the archenteron. Other differences are that the **blastopore** becomes the **mouth** in protostomes while it becomes the **anus** in deuterostomes. Cytoplasmic division in the early embryo of many protostomes shows **spiral cleavage** while many deuterostomes show radial cleavage. With **radial cleavage**, all pairs of daughter cells are similarly oriented which is not the case for spiral cleavage. And many deuterostomes have **indeterminate cleavage** at early stages, meaning that separated cells can give rise to a new embryo. With many protostomes, early cleavage is **determinate**, meaning the developmental fate of cells is set in early cleavage stages.

The best known **protostome phyla** are Nemertia, Mollusca, Annelida, and Arthropoda. **Nemerteans** are elongated, flattened worms called proboscis worms, or ribbon worms. These worms have a defensive and prey-capturing device, the **proboscis**, which is a fluid filled, often barbed, tube that can be forcibly everted from the head to entangle and poison prey.

FLOW CHART OF ANIMAL BODY PLANS

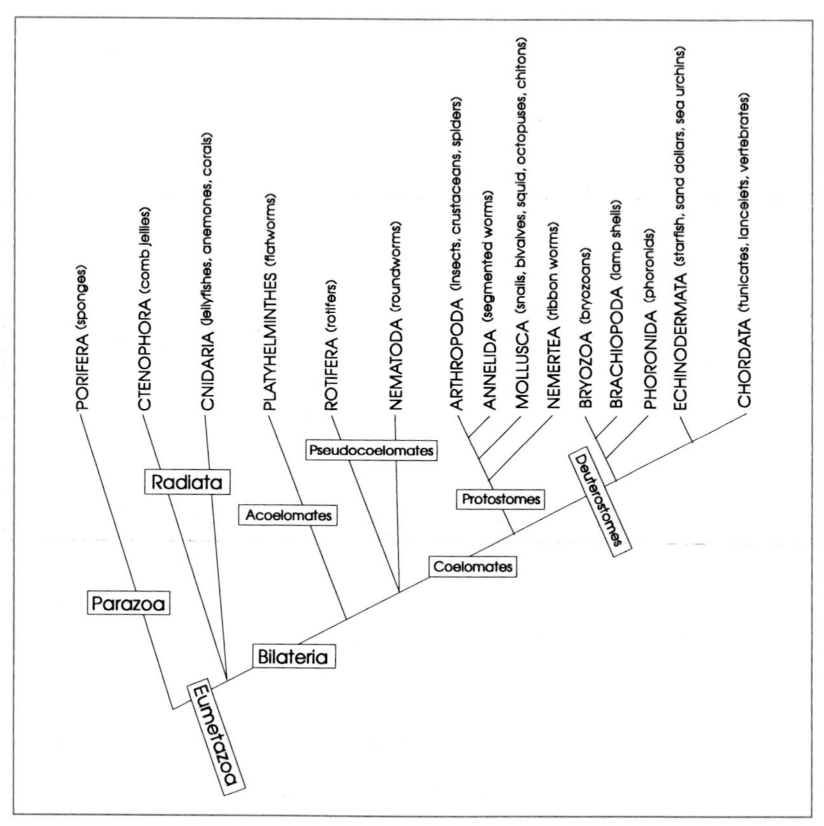

Mollusks are a diverse group of about 100,000 extant species that include, among others, clams, oysters, squids, octopuses, snails, and chitons. They are characterized by a feeding organ, the **radula,** which consists of a cartilaginous base over which slides a membranous belt with chitinous teeth. **Annelids** are the segmented worms: leeches, earthworms, and marine polychaetes. **Metamerism**, or segmentation of the body, into a linear series of similar parts characterizes the annelids. **Arthropods** are the most diverse animal group with about one million species described. Crabs, lobsters and shrimp, horseshoe crabs, scorpions and spiders, ticks and mites, insects, and extinct trilobites are all arthropods. Arthropods are characterized by the retention of some metamerism and by a **chitinous exoskeleton** that covers the entire exterior of the body.

THE MOLLUSCAN RADULA

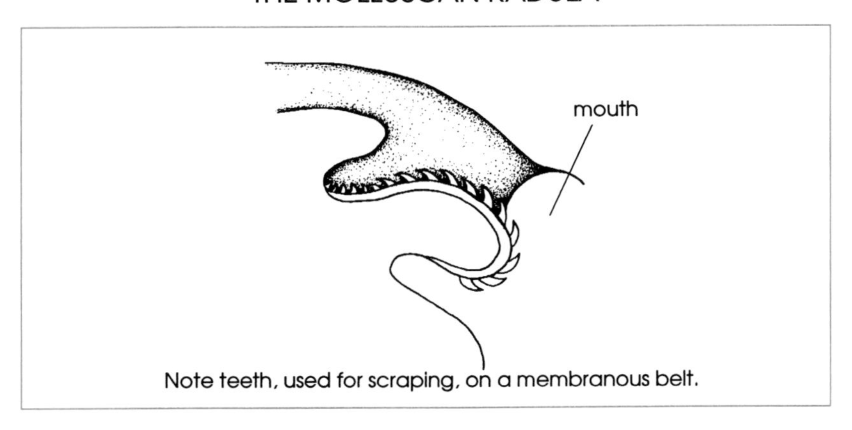

Note teeth, used for scraping, on a membranous belt.

A MOLLUSK

Better known deuterostomes are the Echinodermata, the Chordata, and the lophophorate phyla. **Lophophorate** phyla are characterized by a food-catching organ, the **lophophore**, which is a series of ciliated tentacles, encircling the mouth, that drive a current of water, resulting in the capture of plankton. **Phoronids** are worm-like marine lophophorates that inhabit chitinous tubes. **Bryozoans** are less than 1/2 mm long, mostly marine animals that form encrusting or free-standing, branched colonies. **Brachi-**

opods, or lamp shells, are marine lophophorates that possess a calcareous shell of two valves that resembles a clam shell. **Echinoderms** are non-lophophorate deuterostomes that are among the most familiar of marine organisms: sea stars, brittle stars, sand dollars, sea urchins, sea biscuits, and sea cucumbers. They're characterized by a **water vascular system** which is a series of coelomic canals and surface appendages that may serve locomotory, sensory, and food-collection functions.

MARINE POLCHAETE - AN ANNELID (SEGMENTED WORM)

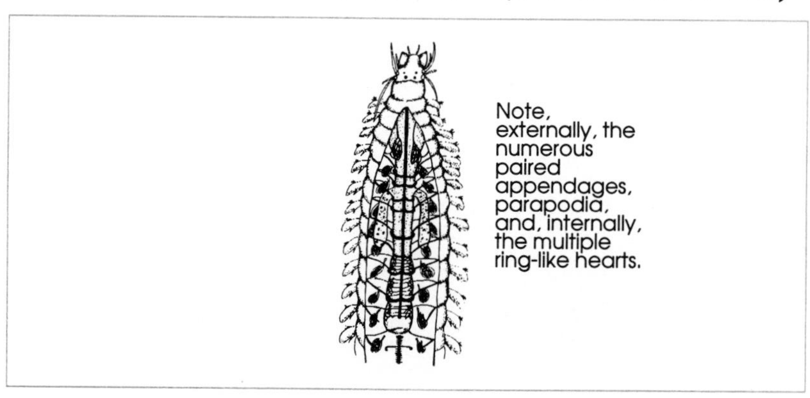

CRUSTACEANS - ARTHROPODS

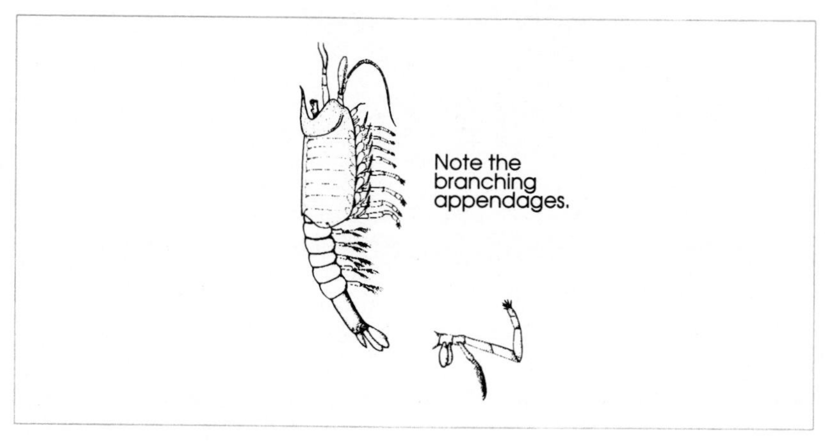

SEA SPIDER (A PYCNOGONID) - AN ARTHROPOD

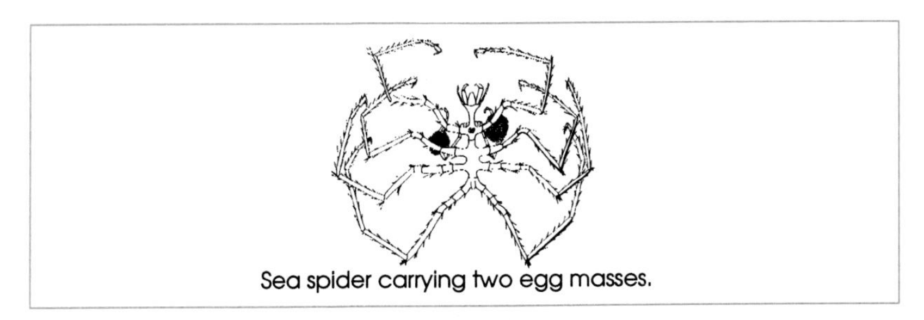
Sea spider carrying two egg masses.

AN INSECT (STINK BUG) - AN ARTHROPOD

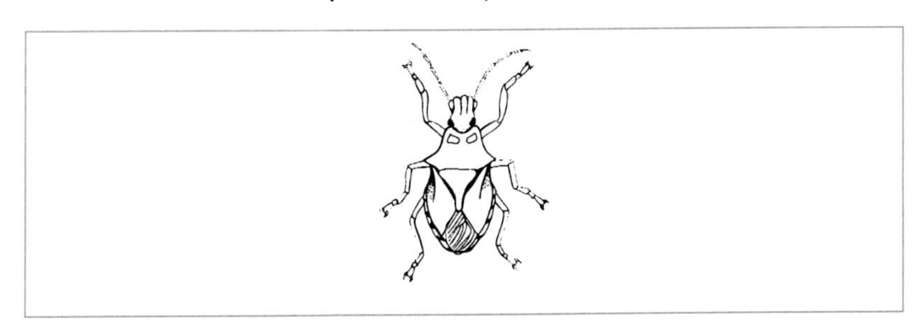

ENTOPROCTS - LOPHOPHORATE-LIKE ANIMALS

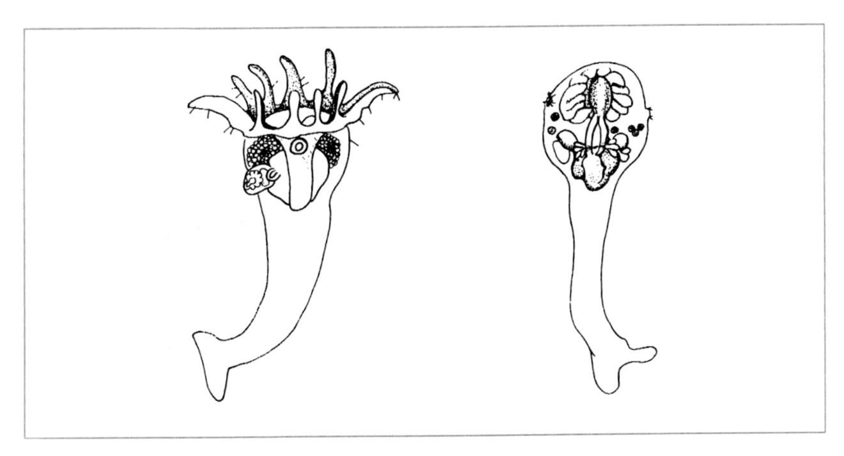

THE SEA STAR - AN ECHINODERM

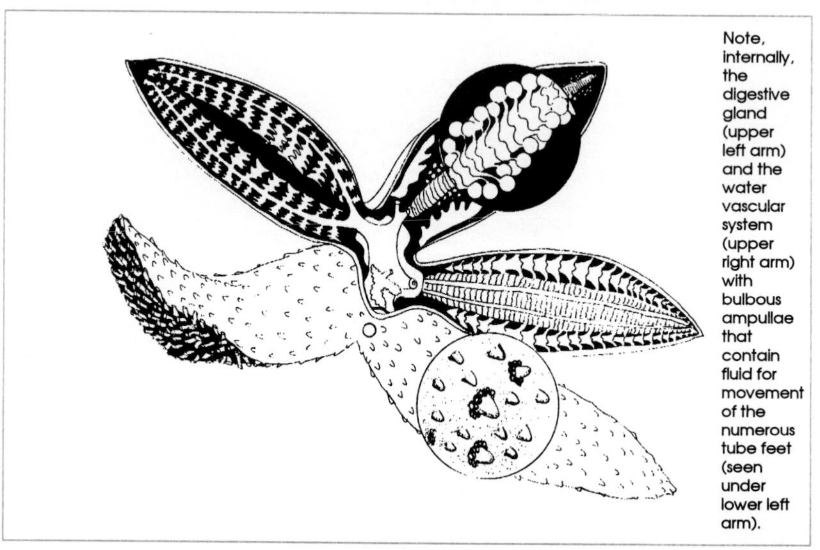

Note, internally, the digestive gland (upper left arm) and the water vascular system (upper right arm) with bulbous ampullae that contain fluid for movement of the numerous tube feet (seen under lower left arm).

Chordates are a very diverse group of non-lophophorate deuterostomes that includes marine **tunicates**, or sea squirts, and vertebrates. **Vertebrates** are fishes, amphibians, reptiles, birds, and mammals. All chordates possess three distinguishing structures at sometime during their life cycle: a **dorsal, hollow nerve chord**, a skeletal element, the rigid, fibrous **notochord**, and **pharyngeal clefts**, also called gill slits.

THE LANCELET - A CHORDATE

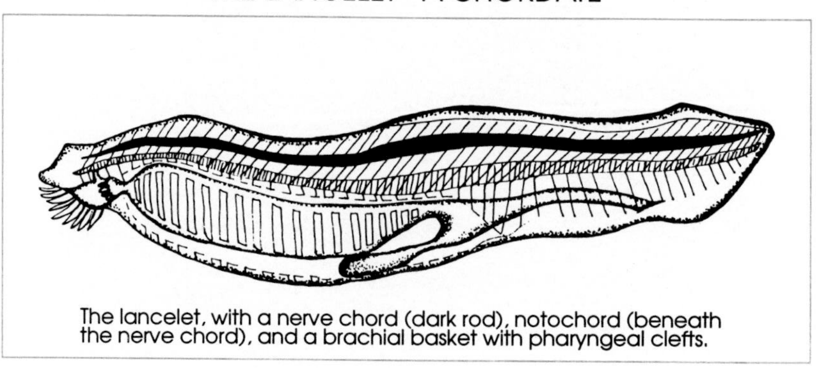

The lancelet, with a nerve chord (dark rod), notochord (beneath the nerve chord), and a brachial basket with pharyngeal clefts.

Now let's **review the key words**: kingdoms Monera, Protista, Fungi, Plantae, and Animalia. Monera is the kingdom that includes the prokaryotes or bacteria. Protistans are unicellular eukaryotes including photosynthetic, autotrophic algae and heterotrophic protozoans. Fungi are heterotrophic organisms characterized by haploid hyphae that grow into a network, the mycelium, the cells of which are bound by chitinous cell walls. Plants are multicellular, photosynthetic autotrophs with cell walls made of cellulose. Animalia is the kingdom of multicellular, heterotrophs that lack cell walls.

CHAPTER 14

HUMAN ORIGINS & SYSTEMATICS

The discipline of **systematics** classifies organisms according to the degree of evolutionary relatedness between them. **Key words** for our discussion of classification are: cladistics, homologous structures, convergent evolution, and levels of classification. **Ancestry** and relationship are inferred by the distribution of traits among the taxa being classified. **Taxa,** taxon for singular, may be individual species, kingdoms, or intermediate categories, such as the phylum. In general, the more closely related two taxa are, the more recently they shared a **common ancestor**. Recently diverged taxa will have many traits that are similar.

Traits that are similar in two taxa because of common ancestry are said to be **homologous**. The limb bones of mammals and reptiles are similar because the ancestry of each group traces back to the same amphibian stock that had similar limb structure. But the eyes of octopuses and vertebrates, although they're of similar design, are not homologous structures because their ancient common ancestor lacked such eyes. Here, similar eye structure represents **convergent evolution**, the independent attainment of similar adaptations.

Homologous structures are useful for classification. **Cladistics** is the branch of systematics that classifies taxa by the presence of shared derived traits. **Shared derived traits** are homologous traits that first evolved in the common ancestor of the group of related taxa. Shared derived traits that evolved more recently can be used to classify organisms at lower taxonomic levels, such as the species. Traits that evolved in the distant past can only be used to organize the evolutionary relationships among higher taxa. The evolutionary "**tree**" or flow diagram that relates one taxon to others is called a **cladogram**. Each branch point, or node, reflects the origin of new traits shared in all taxa above that node.

The levels of classification, from highest to lowest, are: **kingdom**, **phylum**, **class**, **order**, **family**, **genus**, and **species**. Each level can then be divided into sublevels or grouped into super levels. Let's consider some traits that are used to distinguish the human species and its associated higher taxa from all others. For humans we have the kingdom, Animalia, and phylum, Chordata, which have already been discussed. The **subphylum** to which humans belong is the **Vertebrata**, or vertebrates, which are distinguished from other chordates by the development of a **vertebral column** around the notochord. In **jawless fish**, such as lampreys, the vertebral column is poorly developed and the **notochord** persists at maturity. In jawed vertebrates, the notochord gives rise to the central portion of the intervertebral discs but appears to be absent at maturity. The subphylum, Vertebrata, contains eight taxa at the level of **class**; jawless fishes, extinct armored fishes, cartilaginous fishes, bony fishes, amphibians, reptiles, birds, and mammals. Humans belong to the class, **Mammalia**, which is characterized by, among other things, hair, specialized teeth, three inner ear bones, and sweat and milk glands. Within the mammals, humans belong to the **order**, Primates. **Primates** are characterized by a **thumb** and great toe that are **opposable** to the other digits, bone encasing the rear of the eye cavities of the skull, and periodic **uterine bleeding** associated with the female reproductive cycle.

Within the order primates, there are three **suborders**, prosimians, tarsiers, and anthropoids. **Prosimians**, including lemurs and lorises, have moist snouts and **one toilet claw** on each foot. **Tarsiers** have dry noses and **two toilet claws** on each foot. **Anthropoids** have dry noses and only nails on their toes, **no toilet claws**. Anthropoids with the **nares** of the nose widely separated are called **platyrrhines** which belong to the new world monkey superfamily, the **ceboids**. Anthropoids with the nares close together are **catarrhines**. Catarrhines belong to two **superfamilies**, the **cercopithecoids**, or old world monkeys, which have **tails**, and the **hominoids**, or gibbons, great apes, and humans, all of which **lack** a tail.

PRIMATE RELATIONSHIPS

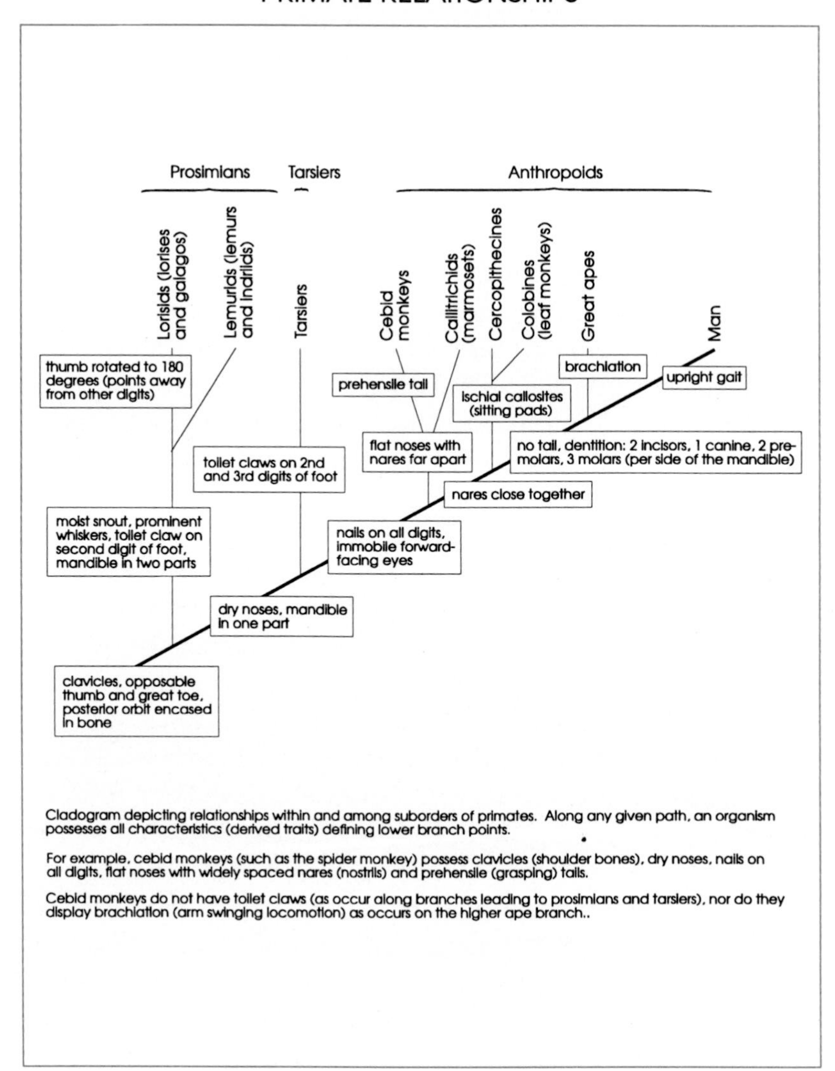

Cladogram depicting relationships within and among suborders of primates. Along any given path, an organism possesses all characteristics (derived traits) defining lower branch points.

For example, cebid monkeys (such as the spider monkey) possess clavicles (shoulder bones), dry noses, nails on all digits, flat noses with widely spaced nares (nostrils) and prehensile (grasping) tails.

Cebid monkeys do not have toilet claws (as occur along branches leading to prosimians and tarsiers), nor do they display brachiation (arm swinging locomotion) as occurs on the higher ape branch..

There are three **families** of hominoids. **Hylobatids**, or gibbons, are small apes with elongated arms for **brachiating**, which is locomotion by arm-swinging through the branches of trees. **Pongids** are the **great apes**, chimpanzees, gorillas, and orangutans. Hominids are humans. **Hominids** evolved about 5.5 million years ago and are characterized by large brains and upright or **bipedal** locomotion. The **oldest** known humans, genus *Australopithecus*, were rather small but had large teeth and powerful jaws. Tool use in hominids dates back two million years. Larger brains and flat faces appearred in the genus *Homo* about two million years ago.

Brain size increased dramatically from the early species, *Homo habilis*, to the later species, *Homo erectus*, and finally to our species, *Homo sapiens*. A now extinct subspecies of humans, the neanderthals, *Homo sapiens neanderthalensis*, actually had larger brains than modern humans.

Now, let's **review the key words**: cladistics, homologous structures, convergent evolution, and levels of classification. Cladistics is classification by the grouping of taxa that share the same evolved or derived traits. Traits or structures that are similar between taxa because the taxa shared a common ancestor with the trait are called homologous structures. Convergent evolution leads to traits that are similar because of similar evolutionary pressures, not common ancestry, and can confuse cladistic analyses. Classification of organisms into the different levels, kingdoms, phyla, classes, orders, families, genera, and species, reflects evolutionary relationships. The lower the level at which organisms are grouped, the more recent their common ancestor.

CHAPTER 15

COMMUNITY STRUCTURE & THE NICHE

Considering there may be 20-50 million extant species prompts questions about how they interact, compete, partition food sources, and adapt to the peculiarities of their respective environments. **Ecology** is the study of how animals interact with, influence, and are influenced by their natural surroundings. The natural surroundings constitute the **environment** which is defined as all the physical and biological factors that impinge on individuals, populations, species, or sets of different species. The **habitat** of an organism is the combination of physical and biological components of the environment that defines where the organism lives. The **niche** of an organism is defined by the organisms adaptations to its environment. Populations of different insect species may occupy the same habitat, openings within the canopy of a jungle, for instance, but will have unique niches because different species are differently adapted. Some species might be adapted to feed on the rotting wood of trees that fall and create gaps while others might be adapted to feed on shade-intolerant saplings that spring up from the forest litter as sunlight floods into newly created gaps.

The habitat has been defined as an organism's **address** while the niche is its **profession**. Three main **axes** or components of the niche are where an animal lives, when it lives there, and how it makes its living. No two species can share the same niche according to the **competitive exclusion principle**. The greater the niche similarity or overlap, the greater the probability that one species will exclude the other from their common habitat by out competing it for one or more essential resources that are in short supply.

Imagine a twisted, knotty stick, encrusted here and there with bark or lichens. The features of the stick are analogized to different possible niches and the whole stick represents all the possible **niche space** available to different organisms in a specific habitat. If there are only a few species in the habitat, they can divide the stick among themselves, each taking a large, separate piece. In this case each species would have a **broad niche**; that is, it would be **generalized** and not specialized with regard to its ability to exploit resources in the environment. As the number of species increases, the stick cannot be divided into large pieces unless species share niche space. But, this means niche overlap increases, leading to the possibility of competitive exclusion. Consequently, we may expect, using the stick analogy, that as the number of species increases, the stick will be broken into smaller pieces, each representing a **narrow niche** to which a particular species is **adaptively specialized**.

The species that occupy a habitat are referred to as a **community**. There is debate about the factors that determine the species composition, or **structure**, of a community. **Competition** among species for limiting resources such as nest sites, refuges from predators, or food has long been held by many to structure communities. The broken stick analogy reflects this school of thought. Here, communities are viewed as assemblages of species that have adaptively **coevolved** to carve up niche space in a manner that reduces competitive overlap. Recently, **predation** has been advanced as a major structuring pressure. Thus, the presence of certain predators in a habitat may limit the species that can survive there. Alternatively, predation could maintain high species diversity by reducing the population sizes of competing species to such an extent that competitive interactions are reduced and exclusion is circumvented.

Competition adversely impacts all species involved because resources in short supply are being shared. Competition may occur in several ways. **Exploitative** or **scramble competition** results from resource depletion. Here, the best competitor exploits the resource more efficiently or at a higher rate. **Pre-emptive competition** reflects a "first-come-first-serve" priority, as in colonization or settlement processes. **Interference competition** entails overt aggression that limits access to an essential resource. Competition, like predation, regulates population growth in a **density-dependent** manner - as population size increases, the negative impact becomes more severe.

Some populations may be regulated by **density-independent** factors such as harsh winter climates that reduce population size by a certain fraction regardless of the actual density. For instance, a blizzard might cull 10% of a bird population whether the population is large or small. There is a debate among ecologists about the relative importance of density independent versus density dependent **regulation**. In more benign tropical environments density dependent regulation may be the more important while in temperate climates with pronounced seasonal changes, density independent regulation may predominate.

CHAPTER 16

WIND, WATER & WEATHER

Climate has a significant impact on the composition of communities within an ecosystem. An **ecosystem** is the soils, all the organisms, and the climate at a particular place and time. Ecosystems may be very broadly defined; for example, a marsh or savanna ecosystem. The global distribution of different ecosystems reflects, in large part, precipitation patterns determined by winds and ocean currents.

Winds and ocean currents are generated by the amount of sunlight energy, or **incident solar radiation**, the earth's rotation, and the tilt of the earth's axis as it orbits the sun. The earth rotates about an **axis** that is approximately perpendicular to the plane of its orbit. Consequently, incident solar radiation is less at higher **latitudes** for two reasons. First, sunlight must penetrate more atmosphere before it reaches the surface, resulting in more reflection away from the surface by atmospheric water. Second, sunlight strikes the surface at smaller angles at higher latitudes resulting in the light being spread out over a greater area just as a flashlight beam changes from a small intense circle of light to a large diffuse triangle as the angle of the flashlight is changed from perpendicular to almost parallel to the wall being illuminated.

Sunlight is more intense, and the temperature warmer, in the **tropics** where solar radiation strikes the earth almost perpendicular to its surface. Heat is retained by the **greenhouse effect**. At night **thermal radiation** from warm surfaces is reflected back to the earth's surface by atmospheric water. Without this reflection of heat back to the earth, night time temperatures would be hundreds of degrees colder than daytime temperatures.

WIND PATTERNS

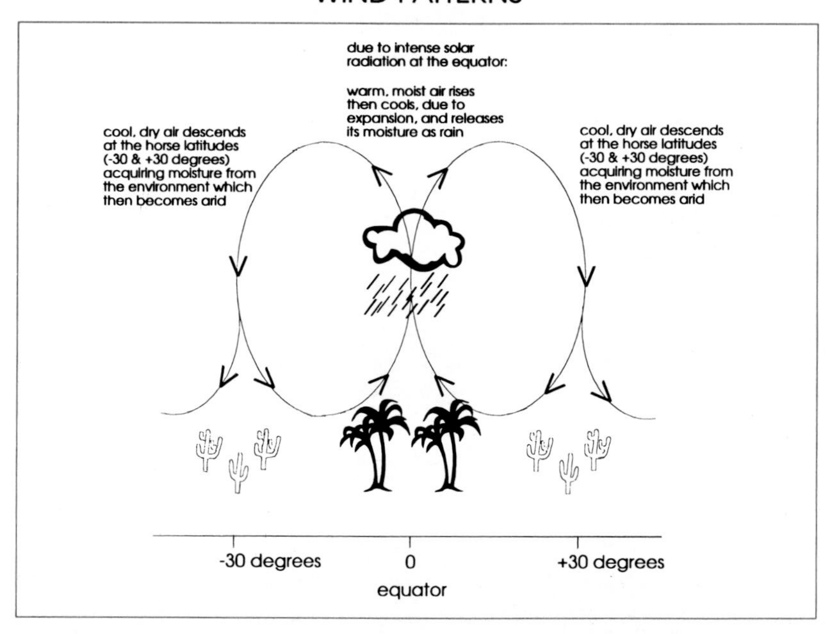

The heating and cooling of air causes **winds**. Intense solar radiation in the tropics warms the air. The warm air expands and rises, moving both northward and southward from the equator. But, the air cools as it rises. The cool air sinks at the **horse latitudes**, 30°N and 30°S. The descending air warms and picks up moisture from its surroundings resulting in deserts at the horse latitudes. The descending air spreads both northward and southward.

Winds, in addition to their north/south component, have an east/west component due to the earth's **rotation**. To understand this, realize that someone at the equator travels farther and therefore faster in the daily west-to-east rotation of the earth about its axis than someone at higher latitudes. This velocity difference causes winds to veer. In the northern hemisphere, winds move north and south from latitude 30°. Those moving north speed up and veer east while those moving south slow down and veer west. In the southern hemisphere, winds from latitude 30° move north and west, slowing, or they move south and east, speeding up. Note that winds from the horse latitudes veer westward as they move toward the equator. These winds are the **easterlies**.

NORTHERN HEMISPHERE SEASONS

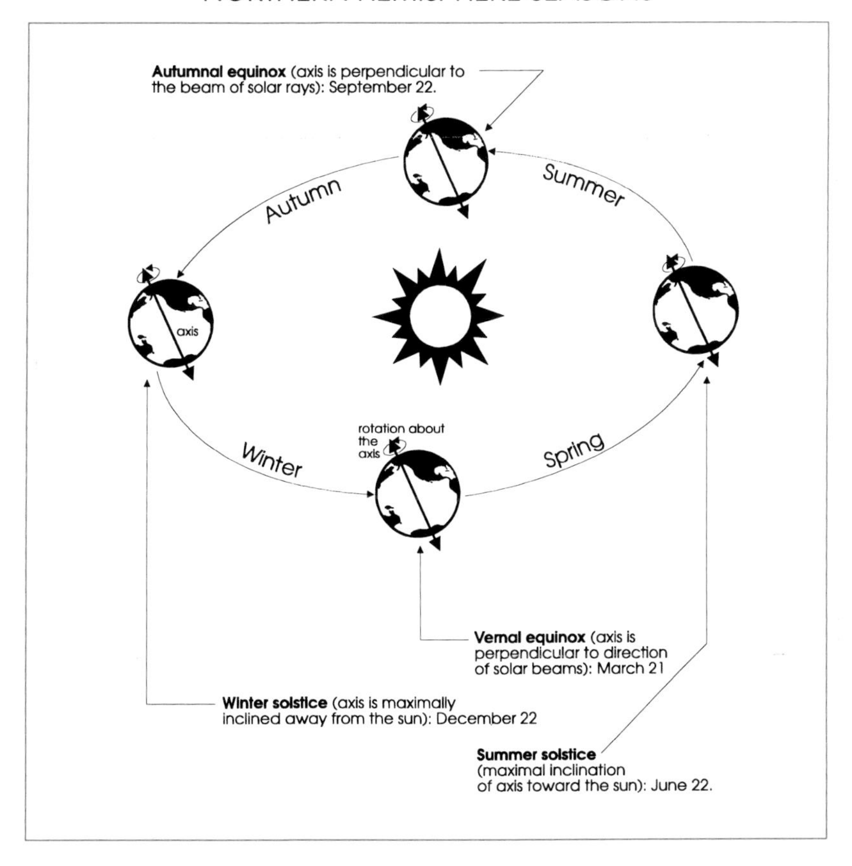

Ocean **currents** arise from the combination of wind pattern and the **coriolis force** which imparts direction to winds and water and arises from the earth's rotation. Currents are clockwise in the northern hemisphere and counter clockwise in the southern hemisphere. Wind and ocean currents vary seasonally because the latitude receiving the most incident solar radiation changes seasonally, from 23°S to 23°N, with changes in the tilt of the earth's axis relative to the sun. **Seasonality** reflects the incline of the earth's axis plus the eliptical shape of the earth's **orbit** that causes the distance from the sun to change. The **vernal equinox**, March 21, is when solar beams are perpendicular to the equator. The **autumnal equinox**, September 22, is when the earth's axis is perpendicular to the plane of its orbit. **Summer solstice**, June 22, and **winter solstice**, December 22, are when the tilt of the earth's axis is maximal, toward and away from the sun, respectively.

RAINFALL PATTERNS

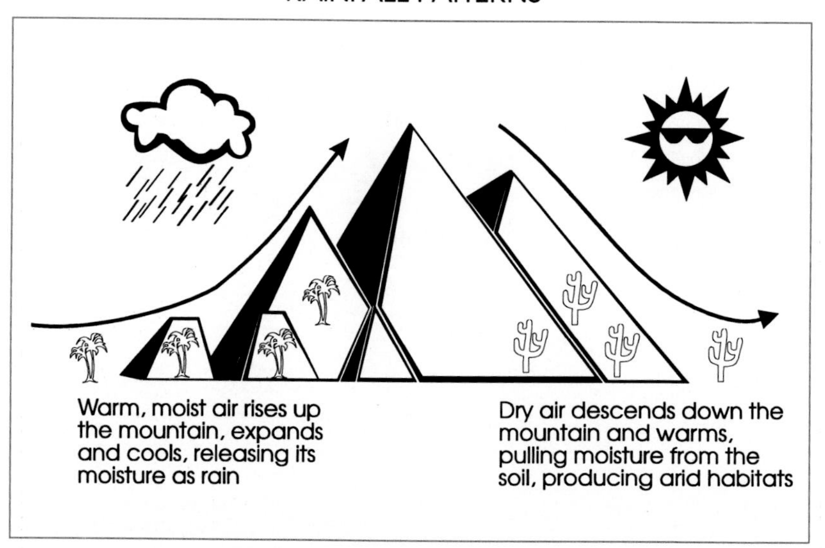

Incident solar radiation and water are the primary factors limiting the productivity of ecosystems. Warm air holds more moisture than cool air. Consequently, the heaviest rains occur near the equator because the air that rises there cools adiabatically as it rises and releases its moisture. **Adiabatic** cooling is due to expansion and not due to heat loss. The horse latitudes receive the least rain because cool air is warming as it descends, increasing its capacity to hold water. Similarly, the windward side of a mountain receives more rain due to the moisture loss from ascending air that cools adiabatically. As the dry air descends on the other side of the mountain it warms and absorbs water causing desication of the terrain.

ADIABATIC COOLING

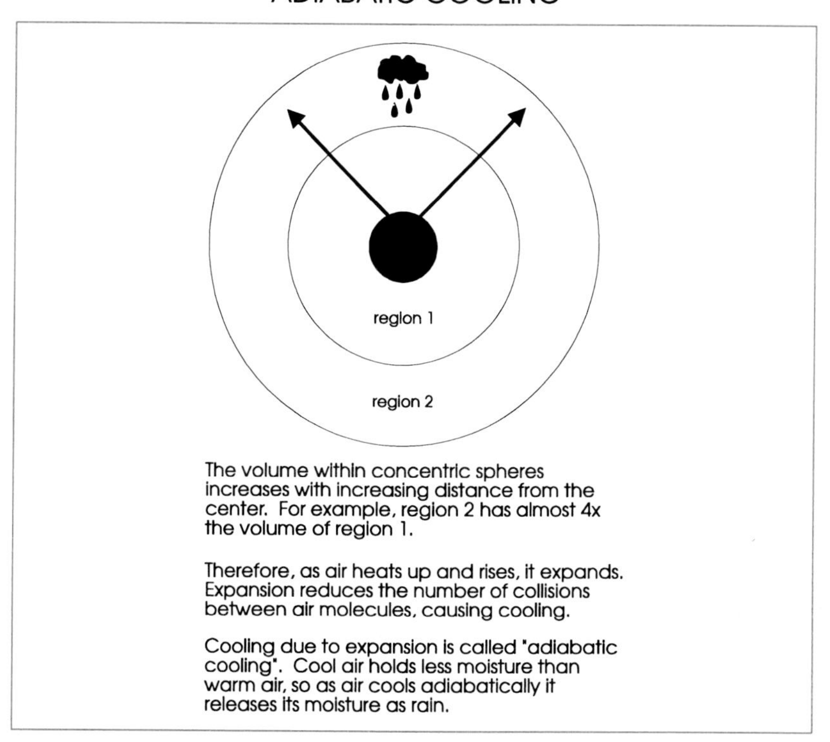

At high latitudes, 30-60°N, the center of **continents** warms faster in summer than does the periphery due to the presence of cool ocean air. Consequently, hot air rises in the center creating a **low pressure** area that draws in water-rich ocean air. As this air warms it rises, cools, and releases water. In the winter, the reverse occurs. The center of a continent cools faster due to the ability of ocean waters to retain heat. Thus, a **high pressure** area develops that sends dry cold air toward the coasts. The collision of polar easterlies and midlatitude westerlies produces rain and snow on **west coasts** during winter. **Eastern coasts**, on the other hand, receive precipitation evenly throughout the year.

CHAPTER 17

ECOSYSTEMS & THEIR PRODUCTIVITY

Precipitation has an enormous impact on plant growth, or productivity, in terrestrial ecosystems. **Primary productivity** is the rate at which sunlight energy is converted into organic matter through photosynthesis by plants, algae, and cyanobacteria. **Gross primary productivity** is the primary productivity for an entire ecosystem. In **terrestrial ecosystems**, gross primary productivity rises with precipitation but levels off at high annual levels, around 200 cm per year. However, the **seasonality of rainfall** is important. For instance, **grasslands** and **chaparrals** result from 50 cm of rain, but the former receives summer rain while the latter receives winter rain. **Evapotranspiration** is the release of water into the atmosphere by evaporation, by transpiration which is water loss from plants, and by animal respiration. **Actual evapotranspiration**, or **AET**, correlates with net annual primary productivity. **Net annual primary productivity** is gross primary productivity minus the energy consumption of the producers: plants, algae, and cyanobacteria. Net primary productivity represents the energy available to consumers such as saprobes and herbivores. In aquatic systems, productivity is often limited by nutrients and light. **Tropical**

ecosystems are more productive than those of temperate zones due to more intense solar radiation and greater rainfall.

Different kinds of ecosystems vary in net annual primary productivity and in their geographic extent. Consequently, very productive ecosystems may contribute relatively little to the world's total net primary productivity. For instance, the net primary productivity of the **open ocean** averages only 125 grams per square meter per year but constitutes 25% of the worlds total productivity while **estuaries** average 2000 grams per square meter per year but contribute only 2.4% of the total. Likewise, **swamps** and **marshes** have net primary productivities of 2000 grams per square meter per year but contribute only 2.4% to the total net primary productivity. Other highly productive ecosystems are: **tropical forests**, 2000 grams per square meter per year; and **temperate forests**, 1300 grams per square meter per year. These ecosystems account for, respectively, 24% and 14% of the world's net primary productivity. Less productive ecosystems are: **savannas**, 700; **lakes**, 500; **continental shelves,** 350; and **desert scrub,** 70 grams per square meter per year.

Together all of these ecosystems represent but a thin, perhaps fragile, film of life on the earth's surface called the **biosphere**. The biosphere scarcely penetrates the earth's crust where thin soils or ocean sediments harbor organisms and is confined to the lower layers of the atmosphere where organisms are shielded from **ionizing solar radiation** and extreme fluctuations in temperature.

This concludes *Genetics, Diversity, and the Biosphere*. The integration of genetics and evolutionary theory into a cohesive, encompassing paradigm for the understanding of biological diversity and ecological interaction continues with as much promise today as when Darwin wrote, at the conclusion of *The Origin of Species*, "It is interesting to contemplate a tangled bank, clothed with many plants of many kinds, with birds singing on the bushes, with various insects flitting about, and with worms crawling through the damp earth, and to reflect that these elaborately constructed forms, so different from each other, and dependent upon each other in so complex a manner, have all been produced by laws acting around us...Thus, from the war of nature, from famine and death, the most exalted object which we are capable of conceiving, namely, the production of higher animals, directly follows. There is grandeur in this view of life, with its several powers, having been originally breathed by the Creator into a few forms or into one; and that, whilst this planet has gone cycling according to the fixed law of gravity, from so simple a beginning endless forms most beautiful and most wonderful have been, and are being evolved."

REVIEW QUESTIONS

Genetics

1. With which base does adenine pair in DNA? (a) adenine (b) cytosine (c) guanine (d) thymine (e) uracil
2. What base is present in DNA but not RNA? (a) adenine (b) cytosine (c) guanine (d) thymine (e) uracil
3. Which of the following bases is a purine? (a) adenine (b) cytosine (c) guanine (d) both a & b (e) both a & c
4. Which of the following terms best describes the orientations of the two strands of a DNA molecule? (a) alpha helix (b) antiparallel (c) double helix (d) right handed helix (e) both b & c
5. What is one of the alternate forms of a gene? (a) repeated simple sequence DNA (b) retrotransposon (c) allele (d) mutation (e) multigene family
6. In which phase of mitosis do sister chromatids move in opposite directions away from the metaphase plate? (a) anaphase (b) interphase (c) metaphase (d) prophase (e) telophase
7. What term refers to the presence of one extra chromosome, relative to what is normal? (a) aneuploidy (b) nondisjunction (c) codominance (d) complement (e) triploidy
8. What is an individual with the genotype **Aa** at the **A** locus? (a) aneuploid (b) heterozygote (c) homozygote (d) homogametic (e) recombinant
9. Which of the following best describes a pair of homologous chromosomes? (a) they both come from the father or both from the mother (b) they carry identical sets of alleles (c) they carry the same set of genes (d) they are joined at the centromere (e) both b & d
10. Which of the following best describes a pair of sister chromatids? (a) they both come from the father or both from the mother (b) they carry identical sets of alleles (c) they carry the same set of genes (d) they are joined at the centromere (e) both b & d

11. What is a germ-line cell? (a) cell that gives rise to zygotes (b) cell type that undergoes mitosis only (c) cell that gives rise to gametes (d) a haploid cell (e) both c & d

12. What does homogametic mean? (a) homozygous condition at all loci within an egg or sperm (b) produced by a mating between phenotypically similar parents (c) possessing two sex chromosomes of the same type (d) produced by inbreeding (e) produces gametes that carry only one type of sex chromosome, the Y chromosome

13. What is the aggregation of centromere-associated proteins that bind spindle microtubules? (a) centriole (b) isochore (c) diakinesis (d) diplotene (e) kinetochore

14. How many hydrogen bonds do guanine and cytosine form between themselves in double strand DNA? (a) 0 (b) 1 (c) 2 (d) 3 (e) 4

15. Who is the father of genetics? (a) Francis Crick (b) Russel Lande (c) Maclyn McCarty (d) Gregor Mendel (e) James Watson

16. The principle of independent assortment was established with breeding experiments on what organism? (a) carnation (b) clover (c) fruit fly (d) human (e) pea

17. What is the involvement of two or more loci in the determination of a single trait? (a) dominance (b) epistasis (c) homeostasis (d) pleiotropy (e) penetrance

18. What is the participation of a locus in the determination of two or more traits? (a) aneuploidy (b) epistasis (c) homeostasis (d) pleiotropy (e) penetrance

19. What process changes the combination of nonalleles on a chromosome? (a) recombination (b) replication slippage (c) nondisjunction (d) chromosomal inversion (e) chromosome condensation

20. Which of the following is not a chemical component of a nucleotide? (a) ribose sugar (b) deoxyribose sugar (c) phosphate group (d) sulphate group (e) both a & d

21. What term refers to genes that lie near one another on the same chromosome?(a) nondisjunction (b) linkage (c) polygeny (d) repetitive DNA (e) synapsis

22. What does the number and type of chromosomes present within a cell constitute? (a) kinetochore (b) genotype (c) karyotype (d) genotypic ratio (e) both c & d

23. What are quantitative traits? (a) those which can be counted in a population (b) those determined by many genes (c) IQ scores (d) traits that are not susceptible to environmental effects (d) both a & d
24. What term refers to the unit of eight histone proteins about which 145 base pairs of DNA are wound? (a) nucleosome (b) ribosome (c) endosome (d) kinetochore (e) genome
25. What process entails the production of two identical diploid nuclei from a single nucleus with duplicated chromosomes? (a) syngamy (b) gametogenesis (c) mitosis (d) meiosis (e) karyogamy
26. What organelle in animal cells gives rise to spindle microtubules? (a) centriole (b) chloroplast (c) lysosome (d) microbody (e) zymogen granule
27. What term refers to all the hereditary material that is characteristic of an organism? (a) DNA (b) gametogenesis (c) genome (d) karyotype (e) diploid
28. What are histones? (a) food particles encapsulated in a lipid bilayer (b) points at which sister chromatids are attached (c) one of the classes of nucleic acid bases (d) proteins around which DNA is wrapped (e) viral enzymes necessary for the production of DNA from RNA
29. What kind of bonds hold together the two strands of a DNA molecule? (a) covalent (b) hydrogen (c) ionic (d) polar covalent (e) either a or c, depending on the bases involved
30. With which base does cytosine bond in DNA? (a) adenine (b) cytosine (c) guanine (d) thymine (e) uracil
31. What term refers to the division of the cytoplasm following mitosis or meiosis? (a) integration (b) cytokinesis (c) kinetochore (d) translation (e) none of these
32. What is a nucleosome? (a) part of the nucleus where ribosomal RNAs are made (b) point of recombination between homologous chromosomes (c) product of nuclear division (d) particle composed of histones (e) point at which spindle microtubules attach to a chromatid
33. Which of the following sets bases are pyrimidines? (a) thymine and adenine (b) guanine and cytosine (c) guanine and adenine (d) uracil and adenine (e) uracil, thymine, and cytosine
34. What molecule is responsible for converting messenger RNA into DNA? (a) histone (b) DNA polymerase (c) ligase (d) connexin (e) reverse transcriptase

35. Which of the following is not true of restriction enzymes? (a) they cleave through double stranded DNA (b) they recognize short, specific, palindromic sequences in DNA (c) they are derived from viruses (d) they are used to produce recombinant chromosomes (e) they are endonucleases

36. Which of the following is not true of commercially available plasmids used in recombinant DNA technology? (a) they are replicated within bacterial cells (b) they carry a cluster of restriction enzyme sites called a "polylinker" or multiple cloning site (c) small pieces of DNA of up to a few thousand base pairs can be subcloned (=inserted) into them (d) they are derived from viruses (e) they carry a gene or genes for resistance to antibiotics

37. What process during meiosis is ultimately responsible for Turner's and Klinefelter's syndromes and for trisomy 21? (a) nondisjunction (b) aneuploidy (c) recombination (d) prophase (e) cytokinesis

38. Which of the following does not occur during mitosis? (a) DNA replication(b) recombination (c) cytokinesis (d) alignment of homologues across the metaphase plate (e) all of these, a-d

39. By what process do researchers amplify specific, small regions of DNA out of a vast genome? (a) subcloning (b) polymerase chain reaction (c) library production (d) ligation (e) DNA sequencing

40. What term describes traits that are determined by numerous loci, each with a small effect on the phenotype? (a) homologous (b) recessive (c) polygenic (d) codominant (e) linked

41. What enzyme catalyzes the formation of covalent bonds within a DNA backbone, as in the production of recombinant molecules? (a) reverse transcriptase (b) DNA polymerase (c) ligase (d) topoisomerase (e) helicase

42. What term refers to DNA plus its associated proteins? (a) deoxyribose (b) chromatin (c) chromosome (d) euchromatin (e) cDNA

43. Which of the following does not occur in prophase I of meiosis? (a) synapsis of homologues (b) condensation of chromosomes (c) recombination (d) DNA replication/chromosome duplication (e) both a & b

44. Which of the following is not characteristic of RNA [as compared to DNA]? (a) uracil (b) single stranded (c) ribose (d) product of transcription (e) presence of introns

45. What process in meiosis changes the combinations of nonalleles along a chromosome? (a) recombination (b) condensation (c) linkage (d) gametogenesis (e) integration
46. What term describes a cell that contains two chromosome complements, one from each parent? (a) haploidy (b) diploidy (c) polyploidy (d) homozygous (e) heterozygous
47. What term applies to chromosomes that do not determine gender? (a) homologues (b) sister chromatids (c) autosomes (d) heterogametic (e) chiasmata
48. What is the blood type of individuals with the following genotype, $\mathbf{I}^A\mathbf{I}^O$? (a) type A (b) type O (c) type AO (d) type AB (e) type Rh^-
49. What term applies to DNA that is produced via reverse transcription from mRNA? (a) nongenic DNA (b) heterologous (c) cDNA (d) Z DNA (e) recombinant DNA
50. What is the late-replicating part of a chromosome to which spindle microtubules attach? (a) centriole (b) telomere (c) euchromatin (d) super helix (e) centromere
51. Many transposable elements of eukaryotes are related to which of the following? (a) plasmids (b) structural genes (c) pseudogenes (d) retroviruses (e) exons
52. What term reflects the merger of population genetics with evolutionary theory? (a) population structure (b) modern synthesis (c) Lamarkism (d) adaptationist program (e) uniformatarianism
53. In what process is the information in mRNA used to specify the production of a polypeptide (protein)? (a) transcription (b) replication (c) duplication (d) translation (e) transformation
54. What is the introduction of foreign DNA into a bacterium via a plasmid? (a) infection (b) transduction (c) transfection (d) transformation (e) translation
55. What is the introduction of foreign DNA into a eukaryote cell via a viral vector? (a) infection (b) transduction (c) transfection (d) transformation (e) translation
56. What is the union of two nuclei, one of the egg and one of the spermatozoan?(a) diploidy (b) homogamy (c) syngamy (d) haploidy (e) synapsis

57. What general term of recombinant DNA technology describes the carriers of foreign DNA? (a) virus (b) vector (c) template (d) gene vehicle (e) plasmid
58. Which pair of the following discovered the structure of DNA: (1) Francis Crick, (2) Gregor Mendel, (3) Margaret Kidwell, (4) Linus Pauling, (5) James Watson? (a) 1 & 5 (b) 1 & 2 (c) 2 & 4 (d) 3 & 5 (e) 4 & 5
59. What is a template? (a) stretch of chromosomal material that has undergone recombination (b) the RNA product of transcription (c) cDNA (d) the plane across which chromatids are arrayed in metaphase (e) strand of DNA used as a guide to the production of a complementary strand
60. Which of the following terms is used synonymously with "gene" but implies as well the position of a gene in a chromosome? (a) linkage (b) structural gene (c) allele (d) locus (e) cDNA
61. **A** is dominant to **a**. What proportion of progeny from the cross, **Aa** x **aa**, are expected to show the recessive phenotype? (a) 100% (b) 75% (c) 50% (d) 25% (e) 0%
62. **A** is dominant to **a** and **B** is dominant to **b**. What proportion of progeny from the cross, **AaBb** x **aabb**, are expected to show the recessive phenotype at each locus? (a) 100% (b) 75% (c) 50% (d) 25% (e) 0%
63. **A** is dominant to **a**. What is the genotype of the mother if none of her 100 progeny show the recessive phenotype that her mate exhibits? (a) **AA** (b) **Aa** (c) **aa** (d) a or b (e) cannot be determined from the information given
64. Two pink-flowered plants were crossed. Of their 140 progeny, 35 were red-, 75 were pink-, and 30 were white-flowered. Which of the following is the most likely genotype of the parents given that capital letters represent dominant alleles? (a) **RR** (b) **Rr** (c) **RrSs** (d) **RRss** (e) **ss**
65. A brown-eyed fruit fly, **Aabb**, mated one with scarlet-colored eyes, **aaBb** (capital letters represent dominant alleles; different letters refer to different, unlinked loci). What proportion of the progeny are expected to have white eyes (=**aabb**)? (a) 16/16 (b) 12/16 (c) 8/16 (d) 4/16 (e) 0/16
66. A bald-headed man, $\mathbf{YX^a}$, and a woman with a full head of hair, whose father was bald, want to have a son. If the allele for baldness, $\mathbf{X^a}$, is recessive to the allele for a full head of hair, $\mathbf{X^A}$, what is the probability that their son would be bald? (a) 16/16 (b) 12/16 (c) 8/16 (d) 4/16 (e) 0/16

67. A population of carnations contains 98 with red flowers (=**RR**), 84 with pink flowers (=**Rr**), and 18 with white flowers (=**rr**). What is the frequency of the allele **R**? (a) 0.1 (b) 0.3 (c) 0.5 (d) 0.7 (e) 0.9
68. In a barnyard 3/4 of the chickens have colored feathers (genotype = **CC** or **Cc**) and 1/4 have white feathers (genotype = **cc**). What proportion of the chickens are heterozygous at the **C** locus if the barnyard population is in Hardy-Weinberg equilibrium? (a) 1.00 (b) 0.75 (c) 0.50 (d) 0.25 (e) 0.00
69. A population of fruit flies is founded by a single red-eyed female (**Bb**) that had mated a brown-eyed male (=**bb**). If her progeny and all other descendants mate at random with regard to eye color, what percentage of the population will have red eyes (=**BB** or **Bb**) in 10 generations? (a) 81-100% (b) 61-80% (c) 41-60% (d) 21-40% (e) 1-20%
70. In humans, **HH** and **Hh** gives dangling ear lobes while **hh** gives attached ear lobes. A random sample of a population contained 252 with dangling and 48 with attached lobes. What is the frequency of the allele **h**? (a) 0.80 (b) 0.60 (c) 0.40 (d) 0.25 (e) 0.16

Evolution

1. What is the absolute fitness of a genotype? (a) the average length of life (b) the average number of offspring produced (c) the frequency of that genotype in the population (d) a measure of the intensity of natural selection (e) 1 - s
2. What is an adaptation? (a) a physiological response to a changing environment (b) a trait evolved under the influence of natural selection (c) a trait increases longevity at the expense of reproductive success (d) a trait used to distinguish one species from another (e) both b & c
3. Which of the following is always necessary if natural selection is to promote evolution? (a) variation in reproductive success that is at least partly inherited (b) small population size (c) variation in survival (d) competition for limited resources (e) both b & c
4. Which of the following is generally believed to be the most common mode of speciation? (a) allopatric speciation (b) reinforcement for ethological isolation (d) species selection (d) sympatric speciation (e) either b or d

5. Which of the following is sympatric speciation believed to require? (a) strong disruptive selection by habitat or microhabitat (b) assortative mating by habitat (c) a period of allopatric isolation (d) postzygotic mating isolating barriers already in place (e) both a & b

6. What does sympatry mean? (a) two populations share overlapping ranges (b) two populations do not overlap in their ranges (c) two populations share the same habitat preferences (d) two populations share the same preferences for mates (e) two populations can coexist without competitive exclusion

7. What is the essence of the biological species concept? (a) speciation occurs in allopatry (b) species are recognizable by their unique morphologies (c) species are reproductively isolated from each other (d) species are continually evolving (e) genetic drift drives the evolution of new species

8. What is comparative biology? (a) testing hypotheses through experimentation (b) generating hypotheses through experimentation (c) dissection at the gross morphological level (d) testing hypotheses by contrasting the structure or habits of different organisms (e) drawing inferences from about an organism based on its behavior, physiology, and anatomy

9. Who wrote *On the Origin of Species by Means of Natural Selection*? (a) C.R. Darwin (b) R.A. Fisher (c) J.B.S. Haldane (d) C. Lyell (e) S. Wright

10. What is the loss of genetic material within a DNA sequence? (a) deletion (b) insertion (c) translocation (d) inversion (e) mismatch repair

11. Which of the following processes can cause the loss or gain of small, tandemly repeated DNA sequences? (a) mismatch repair (b) inversion (c) neutral mutation (d) silent mutation (e) replication slippage

12. What is a single base change in DNA that does not alter the amino acid encoded? (a) neutral mutation (b) frameshift mutation (c) silent mutation (d) transition (e) transversion

13. What is a single base change in DNA that alters the amino acid encoded but does not affect the function of the encoded protein? (a) neutral mutation (b) frameshift mutation (c) silent mutation (d) transition (e) transversion

14. What kind of mutation has occurred if the codon, 5' TAC 3', is changed to 5' TAG 3'. (a) neutral mutation (b) frameshift mutation (c) silent mutation (d) transition (e) transversion

15. What kind of mutation has occurred if the codon, 5' TAC 3', is changed to 5' TA 3'. (a) neutral mutation (b) frameshift mutation (c) silent mutation (d) transition (e) transversion

16. What precise form of isolation exists if two species will interbreed but produce infertile hybrid offspring? (a) premating (b) reproductive (c) postmating (d) ethological (e) allopatric

17. What form of selection occurs on snail color if predatory birds only hunt for the most common color? (a) stabilizing (b) disruptive (c) frequency dependent (d) sexual (e) truncation

18. What is this statement: the rate at which natural selection promotes evolution is proportional to the amount of genetic variation? (a) verbal form of Hardy-Weinberg equation (b) always false (c) fundamental law of natural selection (d) the biological speciation concept (e) the conclusion of J.B.S. Haldane

19. Which of these can impede the rate of adaptation to local conditions? (a) sexual selection (b) gene flow (c) genetic drift (d) all of these, a-c (e) b & c

20. Which of these statements is not true of genetic drift? (a) drift causes allele frequencies to change (b) drift can counter the effect of natural selection (c) drift can facilitate the evolution of new adaptations (d) drift may promote speciation (e) the impact of drift is greater in larger populations

21. What term describes the evolution of dark, cryptic coloration in moths in response to soot pollution? (a) allopatric speciation (b) mimicry (c) uniformitarianism (d) stabilizing selection (e) industrial melanism

22. What is competition between males for access to females? (a) intrasexual selection (b) intersexual selection (c) choice for "good" genes (d) disruptive selection (e) b & c

23. What can cause chromosomal breaks that lead to deletions, inversions, and translocations? (a) point mutations (b) selection for tight linkage and the evolution of super genes (c) transposition [movement] of mobile genetic elements (d) disruptive selection (e) unequal exchange during recombination

24. Who was responsible for the concept of uniformitarianism? (a) C.R. Darwin (b) R.A. Fisher (c) J.B.S. Haldane (d) C. Lyell (e) T.R. Malthus
25. Which of the following best describes the evolution of lungs from the swim bladder of fish? (a) disruptive selection (b) macroevolution (c) microevolution (d) punctuated equilibrium (e) uniformitarianism
26. What is the mode of selection if height in a population evolves from mostly intermediate height to many short, fewer intermediate, and almost no tall individuals? (a) directional selection (b) disruptive selection (c) frequency-dependent selection (d) stabilizing selection (e) truncation selection
27. What is the mode of selection if the distribution of height in a population evolves from mostly intermediate height to many short, few intermediate, and many tall individuals? (a) directional selection (b) disruptive selection (c) frequency-dependent selection (d) stabilizing selection (e) truncation selection
28. What is the mode of selection if the distribution of height in a population evolves from mostly intermediate height to an even greater proportion of intermediate individuals? (a) directional selection (b) disruptive selection (c) frequency-dependent selection (d) stabilizing selection (e) truncation selection
29. The production of sterile ligers and tiglons from crosses between lions and tigers is evidence of what? (a) macroevolution (b) postmating reproductive isolation (c) premating reproductive isolation (d) reinforcement for ethological isolation (e) both a & d
30. Under what conditions would you expect the rate of adaptive evolution to be greatest? (a) weak natural selection and high gene flow rate (b) countervailing [opposing] natural and sexual selection (c) weak natural selection and large population size (d) strong natural selection and large population size (e) high gene flow rate and small population size
31. What is the evidence for punctuated equilibria? (a) abrupt appearance of new species in the fossil record (b) fossil evidence of long-term directional selection (c) fossil evidence of mass extinctions (d) the rise in mammal diversity following the extinction of dinosaurs (e) all of these, a-d
32. Relative to contemporaries in a population, what does natural selection always favor more of? (a) longevity (b) mate selection (c) niche divergence (d) parental care (e) reproductive success

33. What enzyme is critical to the replication of retrovirus [RNA virus]-like mobile genetic elements [TEs]? (a) DNA polymerase (b) reverse transcriptase (c) RNA polymerase (d) topoisomerase (e) both b & c
34. What is the selection coefficient, s_{aa}, if the number of offspring by genotype is: **AA**=10, **Aa**=10, **aa**=9? (a) 1 (b) 0.9 (c) 0.1 (d) 0 (e) -1
35. What is the relative fitness of the **aa** genotype, w_{aa}, if the number of offspring by genotype is: **AA**=10, **Aa**=10, **aa**=9? (a) 1 (b) 0.9 (c) 0.1 (d) 0 (e) -1
36. What is mean population relative fitness if p [=frequency of allele **A**] =0.6 and if the number of offspring by genotype is: **AA**=10, **Aa**=10, **aa**=9? (a) 9.67 (b) 0.967 (c) 15.6 (d) 9.84 (e) 0.984
37. What hypothesis is advanced, by those opposed to adaptationist arguments, to account for macroevolutionary trends? (a) speciation by chromosomal mutation (b) punctuated equilibria (c) reinforcement (d) species selection (e) mass extinction
38. Which of the following is attributed to intersexual selection? (a) bright feathers in birds (b) long canines and horns in mammals (c) species selection (d) sexual dimorphism of the form: males larger than females (e) both b & d
39. Who, along with Charles Darwin, is credited with the concept of natural selection? (a) Ronald Aylmer Fisher (b) Charles Lyell (c) Thomas Malthus (d) Alfred Russel Wallace (e) Sewell Wright
40. What term applies to the concept that the same geological processes operated in the past as in the present which then implies that the earth must be very old [*e.g.*, given the slowness at which mountains arise and are abraded]? (a) continental drift (b) Darwinism (c) sedimentation (d) uniformitarianism (e) volcanism

Diversity

1. Which of the folllowing is never present in members of the kingdom, Monera? (a) nucleus (b) lipid bilayer membrane (c) chlorophyll (d) mitochondria (e) RNA
2. Absence of peptidoglycan in cell walls is characteristic of which of the following bacteria? (a) Archaebacteria (b) enteric bacteria (c) Eubacteria (d) mycoplasmas (e) all of these, a-d

3. Which of the following bacteria produce swamp gas (that has been mistaken for ghosts)? (a) Archaebacteria (b) methanogens (c) cyanobacteria (d) pseudomonads (e) spirochaetes
4. Which of the following are eukaryotes? (a) protistans (b) mycoplasmas (c) acellular slime molds (d) both a & b (e) both a & c
5. Algae are: (a) unicellular (b) autotrophic (c) photosynthetic (d) more related to amoebas than to flowering plants (e) all of these, a-d
6. What are diatoms? (a) drown algae (b) naked amoebas (c) cell surface proteins (d) marine and fresh water algae (e) external protistan skeletons
7. The water vacuole is characteristic of which group of organisms? (a) algae (b) fungi (c) flowering plants (d) echinoderms (e) protozoans
8. What is a rhizopodan? (a) amoeba with broad flowing pseudopods (b) fungus associated with the roots of vascular plants (c) the feeding structure of nemerteans (d) pollen tube cell (e) none of the above
9. Filamentous, tubular hyphae organized into a network, the mycelium, is characteristic of which organisms? (a) archaebacteria (b) bryozoans (c) fungi (d) liverworts (e) Ginkgophyta
10. Chitin is found in the cell walls of which group of organisms? (a) animals (b) arthropods (c) fungi (d) red algae (e) none of these
11. Individuals of which of the following hominid species had the largest average brain size? (a) *Homo habilis* (b) *Australopithecus africanus* (c) *Australopithecus afarensis* (d) *Homo erectus* (e) *Australopithecus boisei*
12. When did the first members of the genus *Homo* evolve [ybp = years before present]? (a) 65,000,000 ybp (b) 12,000,000 ybp (c) 5,500,000 ybp (d) 2,000,000 ybp (e) 200,000 ybp
13. From highest to lowest level of classification, which of the following is in the correct order? (a) phylum order genus species (b) kingdom phylum order class (c) class order genus phylum (d) order genus family species (e) phylum family order genus
14. What is a mycelium composed of cells having two paired, but unfused nuclei? (a) ascogonium (b) conidium (c) dikaryon (d) megaspore (e) microspore
15. What are mutualistic associations between plant roots and fungi for the exchange of nutrients? (a) lichens (b) mycorrhizae (c) petioles (d) phoronids (e) water vascular system

16. Which of the following is not characteristic of sac fungi? (a) asexual spores are produced from hyphal extensions called conidia (b) the fusion of haploid mycelia gives rise to a dikaryotic nodule, the ascogonium (c) karyogamy and meiosis occur within club-like basidia (d) both a & b (e) both a & c

17. Which of the following best describes a cladogram? (a) fresh water bivalve (b) evolutionary "tree" (c) whorl of petals (d) cup fungus fruiting body (e) none of these

18. Which of the following are homologous structures? (a) fish swim bladder and reptile lung (b) bird feather and reptile scale (c) crab "shell" and insect exoskeleton (d) reptile limb bones and mammal limb bones (e) all of these

19. What is suggested by the presence of similar structures in organisms whose last common ancestor lacked the structure? (a) convergent evolution (b) evolutionary loss of the structure in the ancestor (c) erroneous interpretation of ancestry (d) either a or b (e) none of these

20. The water vascular system is characteristic of which group of organisms? (a) echinoderms (b) fungi (c) vascular plants (d) nonvascular plants (e) marine algae

21. Which of the following is most characteristic of annelids? (a) pseudocoelomate (b) chitinous exoskeleton (c) metameric (d) notochord (e) ciliated tentacles organized into a feeding structure, the lophophore

22. Which out of the following group of organisms is the most diverse in terms of the number of species? (a) arthropods (b) mollusks (c) fungi (d) flowering plants (e) chordates

23. What is the most inclusive term for organisms that feed on other organisms? (a) predator (b) saprobe (c) autotroph (d) heterotroph (e) parasite

24. In vascular plants, what tissue conducts water and nutrients from the soil? (a) phloem (b) sieve tube members (c) calyx (d) xylem (e) water vascular system

25. Which of the following features can be used to suggest a closer evolutionary relationship between humans and starfish than humans and insects? (a) the blastopore becomes the anus (b) presence of pharyngeal gill clefts during embryonic development (c) presence of a notochord during embryonic development (d) all of these, a-c (e) none of these

26. Which of the following are not vascular plants? (a) ferns (b) conifers (c) ginkgos (d) whiskferns (e) none of these
27. From first to evolve to last to evolve, which order is correct for these taxa: angiosperms, bacteria, cycads, humans? (a) angiosperms, bacteria, cycads, humans (b) bacteria, cycads, humans, angiosperms (c) bacteria, angiosperms, cycads, humans (d) bacteria, cycads, angiosperms, humans (e) cycads, bacteria, angiosperms, humans
28. From outermost [peripheral] to innermost [central], what is the order of structures of a perfect flower? (a) petals, stamens, sepals, style (b) sepals, petals, stamens, style (c) stamens, sepals, petals, style (d) petals, sepals, stamens, style (e) any of these, a-d
29. From what is the egg of a flowering plant derived? (a) nucellus (b) diploid mother cell within the nucellus (c) one of four haploid megaspores produced from meiosis in the mother cell (d) one of the third generation mitotic products of the one surviving megaspore (e) all of these
30. What is a matured angiosperm ovary and associated flower parts? (a) cotyledon (b) endosperm (c) fruit (d) monocot (e) sporophyte
31. Which of the following is a dicot? (a) palm tree (b) orchid (c) lily (d) walnut tree (e) both a & d
32. Which of the following groups of animals are not bilaterally symmetrical? (a) sponges (b) round worms (c) mollusks (d) jelly fishes (e) both a & d
33. Which embryonic tissue gives rise to the lining of the coelom? (a) endoderm (b) ectoderm (c) mesoderm (d) a or b (e) both a & b
34. Which is true of acoelomates? (a) they have no cavity between the intestine and body wall (b) they lack mesodermally-derived organs (c) flatworms and round worms are acoelomates (d) all of these, a-c (e) none of these
35. Which of the following is true of coelomates? (a) the blastopore may become the mouth (b) the blastopore may become the anus (c) both arthropods and chordates are coelomates (d) all of these, a-c (e) none of these
36. Which of the following correctly matches the animal: bryozoan, echinoderm, mollusk, nemertean with the structure: lophophore, proboscis, radula, water vascular system? (a) bryozoan:lophophore, echinoderm:proboscis, mollusk:radula, nemertean:water vascular system (b) bryozoan:lophophore, echinoderm:radula, mollusk:water vascular system,

nemertean:proboscis (c) bryozoan:radula, echinoderm:water vascular system, mollusk:lophophore, nemertean:proboscis (d) bryozoan:water vascular system, echinoderm:lophophore, mollusk:radula, nemertean:proboscis (e) bryozoan:lophophore, echinoderm:water vascular system, mollusk:radula, nemertean:proboscis
37. Which set of primate taxa includes only those to which humans belong? (a) prosimians, platyrrhines, hominoids (b) tarsiers, anthropoids, pongids (c) anthropoids, catarrhines, hominoids (d) anthropoids, platyrrhines, hominoids (e) ceboids, cercopithecoids, pongids
38. Which of the following sets correctly matches the hominoid: gibbon, gorilla, human to the taxonic family: Hominidae, Hylobatidae, Pongidae? (a) gibbon:Hominidae, gorilla:Hylobatidae, human:Pongidae
(b) gibbon:Hylobatidae, gorilla:Pongidae, human:Hominidae
(c) gibbon:Pongidae, gorilla: Hominidae, human: Hylobatidae
(d) gibbon:Pongidae, gorilla:Pongidae, human:Hominidae
(e) gibbon:Hylobatidae, gorilla:Hominidae, human:Hominidae
39. Which of the following is most closely related to humans? (a) lemur (b) starfish (c) baboon (d) mollusk (e) a & c, equally
40. Which of the following is not a mollusk? (a) snail (b) squid (c) oyster (d) sand dollar (e) chiton

Ecology and the Biosphere

1. What term refers to cooling due to expansion of gas molecules, not loss of heat? (a) adiabatic (b) green house effect (c) density dependence (d) density independence (e) solstice
2. What is the collection of species that occupy a habitat? (a) community (b) ecosystem (c) niche (d) solstice (e) environment
3. Which of the following ecosystems has the highest net primary productivity? (a) grassland (b) temperate forest (c) open ocean (d) marsh (e) chaparral
4. What frequently limits primary productivity in lakes? (a) rainfall (b) sunlight energy (c) nutrients (d) both a & b (e) both b & c
5. What is the combination of biological and physical components of the environment that defines where an organism lives? (a) ecosystem (b) niche (c) habitat (d) community (e) community structure

6. If, by analogy, an organism's habitat is its address then what is the organism's niche? (a) mother (b) profession (c) home (d) language (e) children
7. In which of the following habitats would niche space generally be most finely divided? (a) arctic tundra (b) savannah (c) temperate grassland (d) tropical rain forest (e) both c & d
8. Which of the following best characterizes interference competition over a resource in short supply? (a) overt aggression (b) first-come-first-serve priority (c) altruistic sharing (d) most consumption by the most efficient consumer (e) hording behavior
9. Which of the following is a density-independent regulator of population size? (a) predators (b) intraspecific [within the same species] competition (c) inclement weather (d) interspecific [between species] competition (e) both b & d
10. What entity is composed of all the organisms, climate, and soil and a particular place and time? (a) community (b) habitat (c) environment (d) microenvironment (e) ecosystem
11. At which latitude would the average [yearly] intensity of solar radiation striking the earth be less? (a) 0° (b) 30° N (c) 60° N (d) 45° S (e) 60° S
12. What is the greenhouse effect? (a) depletion of the ozone layer at the poles (b) reflection of thermal radiation by the atmosphere back to the earth's surface (c) release of CO and CO_2 from trees and automobile exhausts (d) fluctuation in temperature between day and night (e) both c & d
13. What causes winds? (a) ocean currents (b) rotation of the earth about its axis (c) solar "winds" of ionizing radiation (d) rain (e) heating and cooling of air
14. What is the net primary productivity [grams/square meter/year], first, of the open ocean, and, second, of a tropical forest? (a) 125, 700 (b) 700, 1200 (c) 700, 125 (d) 2000, 2000 (e) 125, 2000
15. Which of the following represents the energy actually available to consumers? (a) actual evapotranspiration (b) primary productivity (c) gross primary productivity (d) evapotranspiration (e) net primary productivity

16. Which of the following combinations would usually provide for the highest gross primary productivity in a terrestrial ecosystem? (a) high summer, low winter rainfall (b) high winter, low summer rainfall (c) low summer, low winter rainfall (d) low actual evapotranspiration
(e) both a & d

17. At which latitude are arid conditions most likely to prevail? (a) 0°
(b) 30° N (c) 60° N (d) 90° N (e) 45° S

18. What effect does descending air have on the ground beneath it?
(a) buffers it against moisture loss (b) releases rainfall causing AET to increase (c) desiccation (d) cooling winter south easterlies (e) both a & b

19. When are solar beams perpendicular to the earth's equator? (a) noon
(b) vernal equinox (c) summer solstice (d) never because the earth's surface is curved (e) both a & c

20. Which of the following is a possible consequence of interspecific [between different species] competition? (a) extinction of one species
(b) coevolution of narrower niches (c) reduced population sizes
(d) altered vulnerability to predation (e) all of these, a-d

ANSWERS TO REVIEW QUESTIONS

Genetics

1. d	2. d	3. e	4. e	5. c	6. a	7. a	8. b	9. c	10. e
11. c	12. c	13. e	14. d	15. d	16. e	17. b	18. d	19. a	20. d
21. b	22. c	23. b	24. a	25. c	26. a	27. c	28. d	29. b	30. c
31. b	32. d	33. e	34. e	35. c	36. d	37. a	38. e	39. b	40. c
41. d	42. c	43. e	44. e	45. a	46. b	47. c	48. a	49. c	50. e
51. d	52. b	53. d	54. d	55. c	56. c	57. b	58. a	59. e	60. d
61. c	62. d	63. a	64. b	65. d	66. c	67. d	68. c	69. c	70. c

Evolution

1. b	2. b	3. a	4. a	5. e	6. a	7. c	8. d	9. a	10. a
11. e	12. c	13. a	14. e	15. b	16. c	17. c	18. c	19. d	20. e
21. e	22. a	23. c	24. d	25. b	26. a	27. b	28. d	29. b	30. d
31. a	32. e	33. e	34. c	35. b	36. e	37. d	38. a	39. d	40. d

Diversity

1. d	2. a	3. b	4. e	5. e	6. d	7. e	8. a	9. c	10. c
11. d	12. d	13. a	14. c	15. b	16. c	17. b	18. e	19. d	20. a
21. c	22. a	23. d	24. d	25. a	26. e	27. d	28. b	29. e	30. c
31. d	32. e	33. c	34. a	35. d	36. e	37. c	38. b	39. c	40. d

Ecology

1. a	2. a	3. d	4. e	5. c	6. b	7. d	8. a	9. c	10. e
11. a	12. b	13. e	14. e	15. e	16. a	17. b	18. c	19. b	20. e

INDEX